[redacted]

[redacted]

CHEMICAL PIONEERS

CHEMICAL PIONEERS

THE FOUNDERS OF THE AMERICAN CHEMICAL INDUSTRY

By
Williams Haynes

Essay Index Reprint Series

BOOKS FOR LIBRARIES PRESS
FREEPORT, NEW YORK

First Published 1939
Reprinted 1970

STANDARD BOOK NUMBER:
8369-1442-2

LIBRARY OF CONGRESS CATALOG CARD NUMBER:
74-99701

PRINTED IN THE UNITED STATES OF AMERICA

To Dorothy Farrand Haynes
with deep affection and
sincere admiration

CONTENTS

PREFACE

HACKING a wagonroad through the primeval forest, spanning the little streams with neatly laid logs, filling the marshy hollows with stones and brushwood, the pioneer worked his way ever westward. During the past century he crossed the Great Plains; struggled through the heat and dust of the deserts; toiled up over the passes of the Rockies; and in the end reached the Golden Gate. He conquered a continent.

Transferred from mine and farm to the industrial scene, his familiar figure serves as a perfectly fitting pattern for the pioneer American chemical manufacturers.

They, too, were rugged men of boundless courage and undaunted determination. They, also, built new roads and broke virgin soil. They also conquered a continent. With their bare hands they laid down the foundations of that great industry which in our day is fast replacing a mechanical economy based upon coal and iron by a new chemical economy of alloys and plastics and all sorts of new synthetic materials.

Like the pioneer, the individual proprietor has vanished from our American scene. Along the trails blazed by the exploring chemists he built, at the cost of Herculean labor, postroads of chemical commerce which we have transformed into great industrial superhighways. He himself has been succeeded by the corporation executive whose strength and ability are quite as different from his as the skill of the mechanician differs from the talents of the horse wrangler.

It has been fashionable of late bitterly to brand their epoch of business as "the age of fang and claw." Such sarcastic superiority ill-becomes us.

We are the successors of the rugged individualists; the heirs of their industrial conquests. While we can now recognize their faults and errors, we need not belittle their accomplishments. Today especially we might well cherish their good qualities—their daring and perseverance, their spirit of independence, their unflagging industry and brilliant vision—and try to emulate these forgotten virtues.

And so, it has seemed well worth while to tell the life-stories of the men who founded our chemical industry. The companies they built have been merged into our corporations. Often their very names have been lost. To the younger generation of our chemical world they are already becoming legendary figures represented by the name of a subsidiary, of a plant, of some little chemical-making town, or by an oil portrait in the board-room.

Yet far beyond their sentimental and inspirational values, the lives of these chemical pioneers have meaning to us. They epitomize the birth and growth of our chemical industry. They tell the story of our chemical progress, process by process, product through by-product to new product. They reveal the background of our great modern chemical enterprises and why this company is important in alum and why that company keeps out of coaltar derivatives. They explain ancient rivalries and long standing alliances that even today are influences in chemical making-and-selling policies.

. . . .

The writing of this book has been a keen pleasure. Interest in our chemical history has been one of my abiding enthusiasms, and to this has been added a close personal touch. I knew a number of these giants of the past generation as living men. Indeed, I owe several of them—notably, William H. Nichols, Caesar Grasselli, Herbert Dow, Edward Mallinckrodt, George Merck, and John F. Queeny—a lasting debt of gratitude for their encouragement and counsel. Accordingly, I am glad to pay them, and their compeers, this tribute.

Furthermore, the writing of this book has been made far easier than it might have been by the unfailing, generous co-operation of the families and companies of these men. Little of this material has been previously published. Personal records and photographs have been freely placed at my disposal. I have been permitted to examine company archives. I thank all those who have so helped make this book more interesting and authentic.

Originally the chapters that follow were published serially in CHEMICAL INDUSTRIES. This series will be continued, and later a second volume will appear rounding out these brief biographies of the men chiefly responsible for the establishment of a chemical industry in the United States.

WILLIAMS HAYNES

January third, 1939.

JOHN WINTHROP, JR.

1606-1676

THREE hundred years ago—in the autumn of 1635 to be exact—there was set up in Boston a curious chemical establishment which may most rightfully claim to have been the first chemical plant in America. As we know chemical manufacturing, it was hardly a chemical plant at all; but rather a strange combination of druggist's shop, metallurgist's workroom, chemist's laboratory, and alchemist's den. Nevertheless, within it were made experimental batches of alum and saltpetre, and from it radiated a series of primitive industrial enterprises designed to provide the colonists with chemicals, medicines, and gunpowder and to exploit the mineral resources of New England.

Prior to the establishment of this hybrid chemical enterprise, salt had been evaporated from seawater and woodashes had been burned for soap making at both Jamestown and Plymouth, while wine had also been made in the Virginia settlement. However, at a time when leather was home-tanned and cloth home-spun, these were only common household crafts. Bricks and glass and pottery had also been made in America before 1635, but these are process, rather than chemical, industries. To manufacture alum for the home-tanner and saltpetre from which to produce gunpowder are true chemical operations, governed by economic conditions such as control the present day chemical industry, undertaken by a man whose experience and point of view were essentially those of the chemical industrialist.

Chemical manufacturing is a complementary industry: it supports other industries. For chemicals are an unique sort of manufactured raw material. Employed as tools in reactions with other chemicals or upon all sorts of raw materials from hides and fibres to stones and metals they are used to save time or labor, to lower cost, to improve these materials for human use, even to create synthetic materials unknown in Nature. There is no incentive to make chemicals until other industries are ready so to employ them. Fur-skins were an important, early export of the New England colonies, and to manufacture alum for curing them was a perfectly logical enterprise. On the frontier of a new continent gunpowder was a double necessity, for food and for defence; and its most costly, most essential ingredient was saltpetre.

Thus from its very inception, the American chemical industry has promoted national prosperity and national safety. Moreover, from the first it has been marked by a characteristic which today distinguishes this ultra-modern, most scientific of all industries.

JOHN WINTHROP

In a famous definition of chemical industry, John Teeple pointed out that it involves chemical changes resulting in new chemical combinations, and that it handles these changes with chemical intelligence. Unless under chemical control and direction, an operation may be no more a chemical industry than the boiling of an egg by a cook or the slacking of lime by a day laborer, both of which involve chemical changes.

Our first chemical plant was established by a man of real chemical intelligence. Its remarkable mixture of pharmacy, metallurgy, alchemy, and chemistry, was a mirror of the diversified scientific curiosity of the day. In this it clearly reflected the many-sided interests of John Winthrop, Jr.

This father of the American chemical industry might well serve today as the very pattern of a great chemical industrialist. Not only was he a courageous and decisive executive; but he was a man of many affairs, a financier, a soldier, a diplomat. He combined to an exceptional degree that rare combination of scientific interest and economic instinct which always makes the ideal equipment for the active head of any chemical manufacturing enterprise.

John Winthrop, the Younger—so he is distinguished from his father—was an illustrious member of an important family. Son of the great Puritan leader and Massachusetts governor, he was himself the first colonial governor of Connecticut and later one of the Commissioners of the United Colonies of New England. In his chemical and mining operations, he was financially backed by his father. Moreover his son, Fitz-John, carried on these enterprises; accordingly the name of Winthrop has properly a high place among such pioneer chemical families as Harrison, Wilder, du Pont, Innis, Rosengarten, Grasselli, Lewis, and others long and closely identified with the industry.

The Winthrop ancestral home in England was Groton Manor, near Edwardston, in Suffolk, and here both John Winthrop senior and junior were born. The father, after a brief course at Trinity College, Cambridge, entered the law, which he practiced with success, becoming an attorney in the Court of Wards and Liveries and also being engaged in drafting parliamentary bills. Though he continued to make Groton Manor his home, he was a great deal of the time in London until March, 1630, when, having been elected Governor of the Company of the Massachusetts Bay in New England, he sailed for the new world in the *Arabella* in company with a large party of Puritans. An aristocrat in upbringing and taste, strongly conservative in his views, the elder John Winthrop was a mature man when he went through the deep spiritual experience of Puritanism. He quickly became a recognized leader in the Puritan party, and while he frequently opposed the more fanatical doctrines of his fellow colonists, he was the first to defend their political rights from encroachment by the crown and Parliament. His broad views, patience, and courage more than once saved the Massachusetts colony from disaster during its early, perilous years.

John Winthrop

Title pages of some John Winthrop's books with notes on them in his own handwriting.

JOHN WINTHROP

The eldest son followed his father to New England in 1631. At the time John, Jr., was a young man of twenty-five with already a brightly colored background. He had been educated at the Bury St. Edmunds Free Grammar School and at Trinity College, Dublin. Like his father, he did not graduate but studied law at the Inner Temple, London. He volunteered for the expedition, raised by the Duke of Buckingham, to go to the relief of the Protestants of La Rochelle, and after the failure of this ill-fated venture, as was the custom of young gentlemen of his day, he went off for a grand tour. He was absent from England till 1629, travelling chiefly in Italy, Greece, and Asia Minor. Two years later he came to Massachusetts to serve as his father's "assistant," an office he held in 1632, 1635, 1640-41, and again 1644-49.

Young Winthrop landed at Boston the fourth of November, 1631, and with that abundant energy that always distinguished his activities soon won in his own right a prominent position in the colony. Within two years, he was chosen to lead the founding of a new settlement at Agawam (now Ipswich), and in 1634 he was sent back to England to represent the colonists in conference with the English Puritans and to report to the Government upon the threatening encroachments of the Dutch settlers and the Pequot Indians along the foreshore of Connecticut. The following year, 1635, he returned to Massachusetts, commissioned as Governor of Connecticut for one year from Lords Say and Brooke. He sent out a party which built Fort Saybrook at the mouth of the Connecticut River; but except for two short trips, he stayed in Massachusetts and busied himself with his ambitious plans for developing chemical industries based upon the natural resources of the country.

Winthrop returned to England again in 1641 and remained there two years. During this visit he unofficially represented the colonists in a number of negotiations, and he eagerly renewed his friendships among the leading British scientists. In 1645, after his return to Massachusetts, he received the grant of a large acreage in eastern Connecticut and founded there, in 1646, the settlement that became New London. The following year he moved to this new settlement of his and became one of the magistrates of Connecticut. In 1657 he was elected Governor of the Colony of Connecticut and, except for the term of 1658, was annually re-elected until his death in 1676, sixteen years later.

In 1662 he once more visited England upon a diplomatic mission, the success of which was one of the outstanding triumphs of his long career of public service. He secured from the King the famous Connecticut charter which united the colonies of New Haven and Connecticut and which contained strong and dearly cherished civil and political rights to the people of these colonies. Besides serving continuously as Governor of Connecticut, he was also for many years one of the commissioners of the United Colonies of New England.

In the autumn of 1675 King Philip's War was brewing, and a meeting of all the colonial commissioners was called in Boston. Thither John Winthrop hurried in the midst of an exceptionally bitter winter season during which he contracted a fatal illness. He died in Boston, the 6th of April, 1676, two hundred years, to the very day, before the founding of the American Chemical Society.

Very briefly sketched, the swiftly moving, colorful public life of the younger John Winthrop is for our purpose but the background of his chemical interests and his industrial enterprises. From the time of his very first landing at Boston, he was deeply impressed with the desirability of developing the native natural resources and the importance to the colonists of self-sufficiency in iron and gunpowder. He had not been in Massachusetts a year and a half before he sent back to England for laboratory apparatus, books, and chemicals; and what is likely the very first invoice for any chemical importation into America is a statement to him from his English agent, now preserved among the valuable collection of Winthrop papers in the Massachusetts Historical Library:

	l.	s.	d.
Sandiver 2 lbs and soda 8 lbs	0	5	6
Stone blewing 14 lbs	0	10	0
brimstone 1 cwt	1	3	4
copper ¼ cwt	1	10	4
tin ¼ cwt	1	8	0
Canarie seeds 3 pintes	0	0	9
	4	17	11

paid before for the glasses and the charge of packing them and for 3 catalogues of books ..	1	18	5
	6	16	4

received in all	9-12-0
paid in all	6-16-4
I rest indebted to you	2-15-8

Copper, tin, and soda are understandable enough. 'Stone blewing' is obviously blue stone (copper sulfate) and 'brimstone' is certainly sulfur. 'Sandiver' has been explained by that capital chemical antiquarian, Dr. C. A. Browne, as being *suin de verre,* sweat of glass, or the scum which forms on the top of molten glass, used by the alchemists as a constituent of the powder of projection and by the physicians of that time as a cure for gallstones. Canary seeds were also put to medicinal uses in the seventeenth century, being employed, as our grandmothers used flaxseed, for poultices. Undoubtedly, the 'glasses' were crucibles, alembics, etc., for as in those days balances, compasses, magnifying glasses, etc., were known as 'philosophical instruments' so chemical glassware was simply called 'glasses.'

Plainly then, John Winthrop was from the first not only practicing amateur medicine—he did so with such success that his prescriptions soon became famous throughout New England—but he was also making the best effort that the imperfect chemical knowledge of his time permitted to test out his metallurgical and chemical ideas. Simultaneously, he was building up what was to become not only the first, but by far the largest scientific library in the colonies. Two hundred and seventy volumes from this collection of his are now a part of the rich collections of the Society Library of New York. Among them are works on medicine and pharmacy, philosophy, botany, mineralogy, navigation, physics and mathematics, astronomy, and fifty-two chemical books by such ancient authorities as Paracelsus, Dorn, Basil Valentine, Libavius, and Glauber.

It was when Winthrop returned to Massachusetts after his first visit back to England, in the year 1635, that he set to work seriously to execute his ideas of developing chemical resources.

He had for three years explored the native raw materials available for such enterprises. In England he had consulted with the best chemical authorities and conferred with several practical manufacturers. He tested many industrial opportunities. He mined for lead, tin, and copper. He set up works for making salt, glass, and iron. He produced potash, saltpetre, alum, wood pitch and tar, indigo and other natural dye extracts. He was, moreover, the promoter of the first American chemical stock company, and his prospectus—save for the responsibilities placed squarely in the hands of Providence—is quite in the approved Wall Street style:

If any desirous to promote a publique good shall see cause to accomodate that businesse with a stock of 3000£ or 4000£ I shall indeavour (God permitting) to raise such commoditee as may be convenient for returnes, and in particular that staple of saltpeter of which some (blank) of tunes are yearly carried into

England, Holland, Portugall and other parts; and that no adventure of detriment may be to any, doe hereby ingage that the said stock shalbe within (blank) yeares duly repaied to them, with some convenient consideration (if God please to add a blessing to the designe so farre as it be profitably effected); and when it shall appeare demonstratively incouraging, they may, if they please to joyne in the business and to a further proceeding, advance to a stock of 10,000 or 20,000£ or more.

In 1638 he built and operated salt works on the North Shore of Massachusetts Bay, near what is now Beverly. In 1642, in furtherance of his plans to make gunpowder, he secured the passage of an order of the General Court of Massachusetts decreeing that "as will perfect the making of gun-powder, the instrumental means that all nations lay hould on for their preservation . . . every plantation within the Colony shall erect a hous in length about 20 or 30 foote, and 20 foote wide within on half yeare next coming . . . to make saltpeter from urine of men, beastes, goates, hennes, hogs and horses' dung." At the time of this enactment, Winthrop was in England where, among other missions, he was raising the capital and engaging skilled workmen necessary for the establishment of an iron furnace. This was undertaken at Braintree, Massachusetts, in 1644, and the General Court made a grant of 3000 acres to him and his partners for this purpose.

After he moved to Connecticut, he devoted his energies wholeheartedly to the upbuilding of that colony. He established a salt works at New London and operated another iron foundry near New Haven. In 1651 the Connecticut General Assembly issued to him the first monopoly it granted, covering mineral rights so broad and upon terms so generous that had the rocky Connecticut hills been a richer prospecting ground this strenuous and scientifically minded son of the Puritans might well have become a colonial Croesus. His liberal, but not very valuable patent read:

> Whereas in this rocky country, among these mountains and rocky hills, there are probabilities of mines of metals, the discovery of which may be of great advantage to the country in raising a staple commodity; and whereas John Winthrop, Esquire, doth intend to be at charges and adventure for the search and discovery of such mines and minerals: for the encouragement thereof, and of any that shall adventure with the said John Winthrop, Esquire, in the said business, it is therefore ordered by the Court that if the said John Winthrop, Esquire, shall discover, set upon and maintain such mines of lead, copper, or tin, or any minerals, as antimony, vitriol, black lead, allum, stonesalt, salt springs, or any other the like, within this jurisdiction, and shall set up any work for the digging, washing, and melting, or any other operation about the said mines or minerals, as the nature thereof requireth, that then the said John Winthrop, Esquire, his heirs, associates, partners or assigns, shall enjoy forever said mines, with the lands, wood, timber, and water within two or three miles of said mines, for the necessary carrying on of the works and maintaining of the workmen, and provision of coal for the same.

When John Winthrop was in England seeking a charter for his colony, he was elected to the then recently organized Royal Society, and on July 9th, 1662, he read to its distinguished members the first scientific paper ever prepared by an American. His title was "Of the Manner of Making Tar and Pitch in New England," a practical, industrial subject which he handled capably upon the basis of his own experience. He contributed companion papers on the preparation of potashes and black lead. He also exhibited his collection of American mineral and vegetable products and carried on a demonstration of brewing beer from American corn, accompanied by a paper, "Description, Culture, and Use of Maize." After his return to America, he continued to send back papers on a variety of chemical subjects to be read at the meetings of the Royal Society, and till his death he carried on constant, voluminous correspondence with his scientific friends.

He lived just at the time when the revival of scientific knowledge was stirring England. The transition from the alchemy of old to modern chemistry—a change which he appreciated more fully than many of his contemporaries—fascinated him. His restless, pragmatic mind transmuted the new discoveries of chemistry into workaday problems, and we find him continually discussing how some new fact found by experiment might serve some good human use. His alert appreciation of the needs and opportunities of the colonists naturally suggested to him the possibilities of exploiting the mineral resources of New England, and to this end he quite naturally applied his knowledge of chemistry. In spirit, as in fact, he was the founder of the American chemical industry.

His own ventures in this field did not win the material success that his vision, his courage, his persistence merited. He was thwarted by conditions beyond his control. The mineral resources he sought to exploit were not sufficiently rich to justify working. The chemical operations he attempted to carry on had not been sufficiently perfected for economical operation on the edge of the American wilderness. The chemical markets he attempted to supply, though they represented real needs, were not sufficiently large or diversified to support a real chemical production. Accordingly, he failed to establish permanently any profitable chemical or mining enterprise. Nevertheless, his pioneering effort was no sporadic, hit-and-miss "trial." It was a forward-looking, chemically intelligent effort, and John Winthrop, Jr., did more than blaze a chemical trail. In a very real sense he cleared land and built a chemical outpost on the first American industrial frontier.

GEORGE D. ROSENGARTEN

1801-1890

IN the city of Philadelphia, in the year 1822, two ill-mated partners began to manufacture fine chemicals. Both were competent chemists: neither was a good business man. One was a Swiss, speaking only French, while the other was a German who spoke only his native language. Clearly the young firm of Seitler and Zeitler was predestined to disaster.

Within a very few months their difficulties multiplied and entangled into an impassé, and they were forced to call in an arbiter. Their choice of a man for this delicate position was as fortunate as all their previous agreements had been unhappy, for George D. Rosengarten, whom they selected, was ideally fitted for this task. He was a young Westphalian, speaking French and German as well as English, so that, besides being born neutral, he was able at once to comprehend and to interpret the conflicting points of view of the two partners. Moreover, he was a son of bankers trained in banking, and accordingly not only understood the mechanics of accountancy, but also had a detached view

G D Rosengarten

George D. Rosengarten, about the time he entered the chemical industry: a portrait in the possession of the family.

of finances. Both of these attributes were woefully lacking in the equipment of Messrs. Seitler and Zeitler.

All details of the unravelling of this business tangle have unfortunately been lost; but we know that the solution was extremely agreeable to both the embattled partners and most fortunate for the future of the business. His very thorough investigation of the affairs of the firm convinced the young mediator that there was a future for the manufacture of fine chemicals, and he agreed to buy out first one and then the other of the original partners. Accordingly, the firm became Zeitler & Rosengarten, and the little factory on St. John Street was enlarged. The following year (1823) Charles Zeitler retired, and in 1825 George D. Rosengarten, now sole proprietor, moved over to a new building in Adelphi Alley, between Noble and Second Streets. In 1827 the business was moved to Arch Street, just west of Twelfth; in 1829 to the corner of Broad and Vine; and again in 1832 to Vine and Sixteenth Streets where it remained for more than twenty years. These rapid, successive removals were caused by the growth of the business, but in two instances profitable real estate transactions substantially strengthened the resources.

George D. Rosengarten was ever an aggressive manufacturer, and after 1823, when he came into complete control of the enterprise, the number and variety of chemicals produced was increased. During the first ten years of his operations he added quinine sulfate, sulfuric ether, spirits of nitre, ammonia water, acetic ether, and Hoffman's anodyne to the list of C. P. acids and alum originally manufactured. In 1832 the production of morphine salts was begun; the following year piperine was first made; in 1834 strychnine and mercurials; the next year veratain; and 1836 codeine, bismuth, silver salts, and the iodides of lead, iron, mercury and sulfur were added.

In 1840 Mr. Rosengarten took into partnership a young French chemist, N. F. H. Denis, a pupil of the renowned Robiniquet, who had been in his employ since 1835. For thirteen years the firm continued as Rosengarten & Denis, when the junior partner retired to return to France and enjoy the rewards which his technical skill and his partner's business acumen had earned for him. At this time, January 1854, two sons, Samuel G. and Mitchell Rosengarten, were admitted to partnership and the firm name changed to Rosengarten & Sons. The following year, the business was moved to a new plant occupying the entire city block at Seventeenth and Fitzwater Streets, and in 1860 two other sons, Harry B. and Adolph G. Rosengarten, became partners.

In 1879, after fifty-seven years of active leadership, George D. Rosengarten retired, and at this time his youngest son, Frank, was taken into the firm as a partner. Of the sons of the first of the Rosengartens, Adolph, had been killed in a charge, which as senior major in command of the 15th Pennsylvania Cavalry he led at the battle of Murfreesborough, Tenn., December 29, 1862, upon his twenty-fourth birthday; Mitchell

Rosengarten died in 1898 at which time both Samuel and Frank retired. This left only Harry Bennet Rosengarten active in the business of which he became senior partner in association with two of his own sons, George D., named after his grandfather, the founder, and Adolph G., named after his soldier uncle killed in the Civil War. In 1901 the business was incorporated as Rosengarten & Sons, Inc., and in 1905 they bought the plants and business of their ancient and honorable competitor, Powers & Weightman, thus forming the consolidated corporation of Powers-Weightman-Rosengarten Co. In 1921 Harry B. Rosengarten died, the control of the now great manufacturing chemical house passing to his two sons and their younger brothers, Frederic and Joseph, who had in the meanwhile entered the business. In 1927 "P-W-R" consolidated with Merck & Company, and at this time the net assets of the Rosengarten interests (good will and all trade-marks being conveyed without payment) were appraised at over $5,000,000.

In a number of different ways the man who laid down the firm foundation under this great fine chemical manufacturing establishment was an unique figure among our pioneer chemical industrialists. George D. Rosengarten shared their distinguishing traits. He had all the tenacity and self-reliance characteristic of his chemical compatriots; but unlike the majority of them who, even when without the advantages of any schooling in chemistry, were practical makers of chemicals, he was in no sense of the word a 'plant man'. Most of our chemical pioneers learned the lessons of both operation and administration by practical experience; and as self-made men are so prone to do, they relied implicitly upon their own resources and trusted only their own hard-won, first-hand knowledge. He, however, was from the first allied with practical chemists upon whose technical knowledge he relied in the manufacturing end of his business. Moreover, he engaged in business on his own account after a training in banking which had given him, as it were, a second-hand knowledge of industrial and financial administration so that he was enabled to profit by the experience of other men.

Most important of all, however, was a fundamental difference of temperament that set Rosengarten apart from his contemporaries. That sturdy egoism of theirs, so important to their success in the then unmapped fields of industrial chemistry, had the distinct disadvantage of inclining them to domineering habits of thought and action. He, on the other hand, was cooperative by nature, willing to delegate authority, eager to lead rather than to command. Accordingly, while most of our early chemical manufacturers by Herculean labors and often at great personal sacrifice achieved tremendous individual triumphs, they built up great enterprises that were essentially a "one man business"; but George D. Rosengarten created an organization for the production and distribution of fine chemicals. His point of view and his methods were distinctly those of the modern corporation executive. He was thus doubly a pioneer chemical industrialist.

Such a man would naturally be a good competitor, and a rather important by-product of the Rosengarten attitude towards business was the cultivation of a fair and friendly spirit that, in marked distinction to the cut-throat methods of the early heavy chemical producers, characterized the competition between him and John Farr. This is all the more notable because both firms were located in Philadelphia and their products closely paralleled each other.

In those early days—to illustrate the neighborly cooperation that existed—both firms did a big business in quinine salts, and the purchase of cinchona bark, the raw material, was an important task, involving considerable investment and directly affecting important costs, to which both George Rosengarten and John Farr gave their personal attention. This was long before the cinchona tree had been transplanted from the South American jungles to carefully cultivated plantations in the Dutch East Indies, and the wild bark from Peru was shipped to Baltimore as the most convenient American port of entry. Here the incoming stocks were graded and put into warehouses by commission merchants, and here the buyers came for their supplies. In order that two important purchasers might not come into the market at the same time, Rosengarten and Farr were accustomed every two or three months to go alternately to Baltimore. By stage coach, over rough and muddy roads, across ferries rowed by slaves, this was a four days' journey from Philadelphia, often five days in winter and sometimes in summer, if one were very fortunate, three days.

Upon one occasion, after George Rosengarten had started off on one of these periodical trips, John Farr received a letter on the fast packet from London telling him confidentially of a drastic break in the cinchona market. This was indeed important and quite unexpected news which he was perfectly certain his competitor did not know. Now, it would have been quite human of him to have kept this information to himself, to have allowed his rival to stock up in ignorance of the lower prices quoted in what was then the world market headquarters, to have himself purchased his next lot of this important raw material at a price that would give him a clear advantage in costs. Most competitors would have gleefully accepted this break of fortune and thanked their lucky stars. Not so, John Farr.

It was not possible for him to pick up a long distance telephone. He could not even send a telegram. He had to hire a special post chaise, arrange for relays of fresh horses, hurry home, pack his valise, and set forth himself on a long, bumpy, sleepless drive in order to warn his competitor. All that he did.

The man who could inspire such sincere friendship in a keen business rival must have been himself a good friend, and indeed the old city of Philadelphia has seldom had a more universally beloved and respected citizen than George D. Rosengarten. It has been said of him that "he was the worthy successor of Benjamin Franklin and Stephen Girard." He lived to but a single year short of ninety, having come to Philadelphia when he was eighteen, and throughout his long and active career he resided in that city. Though no politician, he was for many years a devoted servant of the city's public and charitable corporations. He was a director of a savings institution and of the Pennsylvania Railroad, and at the time of his death in 1890, the oldest member of the Franklin Institute, the German Society, and the Philadelphia Club, to which he was elected in 1844 when the club-house was situated at 255 Walnut Street.

George David Rosengarten was born in Cassel, Hessen, Germany, on June 30, 1801. He sprang from an ancient and influential family which for several generations had been bankers to the Electors. Always wealthy, the family had very greatly increased their fortune during the era of prosperity brought to Germany at the time of the American War of Independence; but their resources had been completely dissipated by the wars with Napoleon and by the unfortunate reign of his brother, Jerome, in Cassel, as King of Westphalia. The boy, having received a splendid education from the most competent teachers in Cassel, was forced by the political situation and the reduced circumstances of his family, to seek his fortune abroad. He went first to Holland, the home of his mother, for a year's apprenticeship in finance with the great banking house of Hope. Having completed this course, he decided to come to America, and armed with letters of introduction from Hope to leading merchants in Philadelphia, Baltimore, and New York, he landed in Philadelphia in 1819. Immediately he found employment with a firm of wool merchants and shortly afterwards engaged in consulting work as an expert accountant. It was in this professional capacity that he was first brought in touch with the affairs of Seitler & Zeitler and so became interested in chemical manufacturing.

In 1826, on New Year's Eve, he married Elizabeth Bennet, a daughter of Jacob and Henrietta (Bacher) Bennet, born in Hamburg, November 11, 1809, who, when she was ten years old, had come to Philadelphia with her parents. Her mother, a native of Potsdam, was closely related to the Mendelssohn and Hertz families in Hamburg where her father's family had been established for several generations.

Mr. and Mrs. Rosengarten were a most companionable couple, sharing similar tastes in the arts, a deep enjoyment of good music, and a fondness for whist and bezique. These card games, especially as he grew older, were one of his chief means of relaxation. He was a most skillful whist player, and the opportunity "for a good game of whist" was one of the attractions that took him for many seasons to fashionable Saratoga Springs for his summer vacations. Mrs. Rosengarten was a charming hostess and a devoted mother to her large family. They lived together for fifty-nine years (she died at Newport, R. I., July 25, 1885), a complete and pretty picture of the gentleman of the old school and his beloved lady.

The Rosengartens had nine children, six sons and three daughters, all of whom, save a daughter who died when nine years old, they raised to man-and-womanhood. The care with which all of these sons and daughters were educated was conspicuous. Appreciating, as George D. Rosengarten did, the importance of technical training, those boys who later came into the chemical firm were all given the opportunity for a thorough schooling in chemistry. Five of them came into the business with him while the sixth entered the legal profession. The notable success in life achieved by all of these sons is not the least important tribute that may be paid to the founder of the Rosengarten dynasty.

Samuel G. Rosengarten, the eldest son, was born November 8, 1827, and after training in a private school, entered the University of Pennsylvania from which he was graduated with the degree of Bachelor of Arts in 1845, being at the time seventeen years old. To complete his chemical education he sailed on a clipper ship to study in the laboratory of the illustrious Baron von Liebig. Later he enrolled at the University of Paris under Professor Rose. He was living in the French capital during the Revolution of 1848 and not only witnessed the sacking of the Tuileries Palace, but went through the thrilling experience of being impressed into service as a soldier in the National Guard from which uncongenial duty he was promptly released on proving his American citizenship. He returned to Philadelphia on one of the first steamships to cross the Atlantic and immediately went to work in his father's chemical factory. In 1853, at the time the French chemist, Denis, retired from the firm, he was admitted into partnership. In 1898 he himself retired from active business and he died, unmarried, May 15, 1908.

The second son, Mitchell G. Rosengarten, who was taken into the firm at the same time as his older brother, and who died in 1898, at which time this same brother retired from the firm, was a thoroughly American-trained chemist. He attended James School, a private institution, at Eleventh and Market Streets, and graduated at nineteen from the Philadelphia College of Pharmacy. His brother's exciting experiences in Paris, thrilling enough to a young man, but most disturbing to his fond parents, suggested that this younger brother had best be kept at home, so his chemical education was completed by a course in the laboratory of Professors Martin Boye and James C. Booth, then located at the corner of Seventh and Market Streets. Professor Boye enjoyed the reputation of being at this time the foremost chemical teacher in the United States, and young Mitchell Rosengarten, on completing the lessons under him, went straight to his father's laboratories. He married, on August 11, 1868, a great granddaughter of the Irish patriot Samuel Neilson, Emily Huntsman, and they had six children, three of them sons, but none of them ever became identified with the family's chemical interests.

Adolph G. Rosengarten, the Civil War hero, was the fifth son and the seventh child of George D. Rosengarten, born December 29, 1838 and killed in action near Murfreesborough, Tenn., December 29, 1862. Until he was sixteen years old he studied in private schools in Philadelphia and was then sent to Germany to complete his chemical education. He was at once a brilliant student and keen sportsman, and on entering his father's business plunged into his work with characteristic vigor. At the same time, as in Germany he had taken long walking trips through the Hartz Mountains, the Black Forest, along the Rhine, over the Alps, and down into Lombardy, so in Philadelphia he was soon a leading spirit on the cricket field and in boating, swimming, and skating on the Schuylkill. Upon the outbreak of the war between the states all his energies were devoted to the Union cause, and he was one of the organizers of the famous "Anderson Troop," an exceptional body of picked men who rendered notable service in the campaigns in Kentucky and Tennessee under Generals Buell and Rosencrans. Young Rosengarten was mustered into service as orderly sergeant of this famous troop and in a year had been promoted to Senior Major of the regiment that had been added to the original troop. Early in December, 1862, this regiment, with many recruits, was ordered from Louisville to Nashville and thence proceeded to the front where on Christmas Day they were brigaded in a cavalry division led by General Stanley. In his report of the ensuing engagement, that officer wrote: "On the 29th the Anderson Cavalry behaved most gallantly, pushing a full charge upon the enemy for six miles; having dispersed the Rebel Cavalry, they fell upon two regiments of Rebel Infantry in ambush, and after a gallant struggle, were compelled to retire, with the loss of Major Rosengarten and six men killed."

The youngest son of George D. Rosengarten, Frank H. Rosengarten, was born in Philadelphia, May 6, 1843. As a boy he attended Faires Classical Institute, and during the Civil War served in the Landis' Battery. After the war he went to Germany where for two years he studied chemistry, returning to work for Rosengarten & Sons, being admitted to partnership in 1879. In 1923 he died. He married Mary D. Richardson, author of "Eight Journeys Abroad," and they had two sons, one of whom became a lawyer, the other a civil engineer.

Although the fourth son and sixth child of the first of the Rosengartens, Harry Bennet Rosengarten has been purposely kept till last because he was the real successor of George D. Rosengarten in the chemical business, and it is through his sons that the succession was carried on for the third generation until the business was consolidated with Merck & Company in 1927.

Harry Bennet Rosengarten was curiously enough the only son of George D. Rosengarten who did not receive a highly specialized training in chemistry. After a very thorough preliminary education in the small private school conducted by Dr. S. B. Wylie Crawford, he went, at his own suggestion, straight into the accounting department of his father's firm. In doing so he followed directly in his illustrious sire's footsteps, and like him, he carefully familiarized himself with every branch of all the business departments of the manufacturing chemical business. He was admitted as partner in 1860, and finally, in 1898, upon the death of one brother and the retirement of two others, became the senior partner in association with two of his own sons, George D., 2nd, and Adolph G. When the business was incorporated in 1901 he became president, and he held this same office in the consolidated Powers-Weightman-Rosengarten Company, after the purchase of Powers & Weightman in 1905, until his death, February 19, 1921.

The sixty-eight years that cover Harry B. Rosengarten's active business career witnessed the great industrial expansion of this country and the rapid development of the sciences of pharmacy and chemistry. He led the company through two serious financial panics and two wars that brought abnormal growth to its manufacturing facilities. Throughout these long years of great change he carried on the Rosengarten tradition in the spirit of his father, and he died, full of years and honors, the dean of the American chemical industry.

He had married Clara Johanna Knorr in 1868, and they had five sons and two daughters. Four of these sons—George D., Adolph G., Joseph G., and Frederic—comprised the third generation of the Rosengarten dynasty.

The oldest, George D. Rosengarten, named after the grandfather, was educated at the University of Pennsylvania and took a Ph.D. in chemistry at the University of Jena. He became a partner in Rosengarten & Sons in 1898, and when the firm was incorporated, became vice-president, continuing in this office in the consolidated Powers-Weightman-Rosengarten Company and retiring at the time of the merger with Merck. He made a distinguished record of public service, having served as president both of the American Chemical Society and the Institute of Chemical Engineers; a trustee of the Philadelphia College of Pharmacy and of the Franklin Institute, a member of the Revision Committee of the U. S. Pharmacopeia, and a director of the Philadelphia National Bank.

The second son, Adolph G. Rosengarten, was also graduated from the University of Pennsylvania. He entered the firm of Rosengarten & Sons in 1892 and became a partner in 1898. He was secretary and treasurer after incorporation and continued as treasurer after the consolidation with Powers & Weightman until 1921 when he succeeded his father as president. He retired at the time of the merger with Merck & Company. He saw active service in Porto Rico with the First Troop, Philadelphia City Cavalry, during the Spanish War, and was Chief of the Miscellaneous Chemicals Section, Chemical Division of the War In-

Samuel Rosengarten

Mitchell Rosengarten

Adolph G. Rosengarten

Harry B. Rosengarten

Frank H. Rosengarten

ROSENGARTEN & SONS,

MANUFACTURING CHEMISTS,

South West Corner of Seventeenth and Fitzwater Streets,

G. D. ROSENGARTEN.
S. G. ROSENGARTEN.
M. G. ROSENGARTEN.

PHILADELPHIA.

☞ *The Spruce and Pine Street Omnibuses, from the Exchange, pass Seventeenth Street a few squares from the Laboratory.* ☜

CONFIDENTIAL. **1856.** **PRICES SUBJECT TO FLUCTUATION.**

ACID ACETIC, No. 8, American, in carboys,	lb.			CINCHONINE, in 1 oz. vials,	oz.		
" " " English, " "	"			" SULPHATE, in 1 oz. vials,	"		
" " Glacial,	"			COLLODION,	lb.		
" CITRIC,	"			" in 1 oz. vials,	doz.		
" GALLIC, in 1 oz. vials,	oz.			CONIA, in 1/8 oz. vials,	oz.		
" MURIATIC, chem. pure,	lb.			COPPER, Black Oxide,	"		
" NIT[illegible] " "	"			IODIDE[illegible]	"		
" "	"						

Clipping from a price pamphlet of Rosengarten & Sons, dated 1856. This price list listed 175 or more chemicals offered by the firm at that time.

dustries Board, during the World War. He is a director of two banks, two insurance companies, and of the Pennsylvania Salt Manufacturing Co., a trustee of the University of Pennsylvania and president of the Lankenau Hospital.

The two younger sons, Joseph G. and Frederic Rosengarten, are graduates of Princeton, and they both continued active in business after the consolidation with Merck & Company, Frederic becoming chairman of the board at the time of the merger. He, like his older brothers, serves on the boards of several notable public and charitable institutions, being a trustee of Franklin Institute and of the Wistar Institute of Anatomy and Biology, and the president of the board of the Chestnut Hill Hospital.

So concentrated a chronicle of genealogy and honors is needful to record within reasonable limits the distinguished record of this American chemical family. They stand foremost in age, in leadership in both the field of chemical accomplishment, and the broader field of public service, among all of the family groups long and intimately identified with chemical industry in the United States. As they can be justly proud of the family tradition handed down by the first of the Rosengartens, he, too, might well take pride not only in the distinctions which his descendants have won in many fields, but also in the brilliant manner in which they have preserved the honorable ideals and chivalrous spirit of that tradition.

MARTIN KALBFLEISCH

1804-1873

STRONG family tradition, vigorously impressed upon growing youth, must in later years either become the dominating rule of life or else be broken with completely. Such potent influences brook no compromise, and in all the annals of the American chemical industry the force of a virile chemical manufacturing tradition was never more tellingly exemplified than in the family of Martin Kalbfleisch.

The Kalbfleisch home was in the most literal sense an adjunct of the Kalbfleisch chemical works. While yet in school, the Kalbfleisch boys alternately shared plant inspection rounds at midnight and at two in the morning and reported to their father at seven o'clock breakfast the readings of half a dozen important gauges and meters. Their regular studies were supplemented by chemistry lessons administered unsparingly by a humorless Scotch foreman, who before entering the Kalbfleisch employ had been a professor of the science at Edinburgh University.

Yet tradition does not portray the founder of the family as a dour parental martinet. Strict in his ideals of duty Martin Kalbfleisch undoubtedly was, and it was certainly his life-long habit to tend whatever business he had in hand with absorbing concentration; but he was by nature a rather jolly, distinctly neighborly man. Very stern, yet very kindly, he was himself the product of his own upbringing, a younger son of a substantial burgher family of Flushing, Netherlands. In him were early ingrained the traditional virtues of the good Hollander: industry, thrift, and scrupulous honesty.

Martin Kalbfleisch was born in Flushing on February 8th, 1804; but, save that he received an exceptionally good preliminary training in the private schools of his native town, nothing is known of his boyhood. The first event in his life that he himself considered worthy of record was the important decision to seek his fortune in the Dutch East Indies.

Accordingly, at the age of eighteen he took passage for Sumatra in the stout, three-masted windjammer "Ellen Douglass" whose home port was Salem, Mass., and whose shrewd New England captain was destined by Fate to turn his ambitious, but frankly opportunist plans from the Dutch colonies to the United States.

Late in 1822 the "Ellen Douglass" had scarcely dropped anchor in the mountain-rimmed harbor of Padang when she hastily put about and scudded out again into the South Seas. A plague of Asiatic cholera was raging fearfully all along the Sumatra coast, and though the adventurous Yankee trader plied among other islands, putting in to bargain calico and tenpenny nails for copra and pearls, nevertheless young Kalbfleisch resisted the blandishments of the Orient and returned with her to Europe. It is likely that already he had determined to settle in America. When the "Ellen Douglass" returned to Antwerp she was promptly sold, and her Yankee skipper took the young Hollander into a trading partnership with him. Off they posted by stagecoach to Paris, and here Martin Kalbfleisch lived during the next four years.

Although this commercial venture did not prove profitable, nevertheless his stay in Paris had important results. While there his interest was directed towards chemistry, and to his determination to seek his fortune in America was added the definite objective of becoming a manufacturer of chemicals. To this end he attended the chemical lectures at the Sorbonne, then the foremost school of chemistry in the world. While studying there he met a young English girl, Elizabeth Harvey of Southampton whom he promptly married; and as he said later, thoroughly disproved the old proverb about marrying in haste, for he never regretted it afterwards. A year later, after the birth of their first child, he came to America.

In 1826 the opportunities for a young chemist, however well trained and ambitious he might be, did not appear on the surface to be particularly brilliant in the United States. Timber and turpentine, cotton and corn, cattle and fish, land and minerals—almost anything—

The Martin Kalbfleisch's Sons Works operations as illustrated in an article in "The Scientific American," November 20, 1880.

seemed to promise greater rewards than chemicals. Against the golden opportunities that agriculture and commerce dangled in front of any energetic youth, the prospects in industry appeared uninteresting and uncertain. Only a very discerning young man could foresee the economic forces about to burst forth in all sorts of manufacturing activity: a prophetic vision was necessary to anticipate the future demand for all sorts of chemicals.

At that time the meager supplies of chemicals needed by our truly infant industries were practically all shipped from England. Although it seemed most likely that it would be a long, long time before a domestic industrial market would develop that could support a production in the United States of any but the most costly and difficultly transported chemicals, yet already a few pioneers were laying the foundations of our chemical industry. In Philadelphia, then the unqualified center of chemical activity, John Harrison and Charles Lennig were making sulfuric acid; Samuel Wetherill was producing paint pigments; Farr & Kunzi were manufacturing fine chemicals. Just about the time that young Kalbfleisch reached New York another young Hollander, George Rosengarten, was straightening out the tangled finances of a couple of young Swiss chemists, Zeitler and Seitler, who in 1822 had begun to manufacture mercurials and other medicinal chemicals.

In New York a group of merchants, headed by John C. Morrison, James Jenkins, Geradus Post, and Charles G. Haynes, had recently organized a stock company to manufacture sulfuric acid, blue vitriol, lead acetate, and other chemicals. They had met with initial success and were planning to expand their plant, adding to their list of products. They were handicapped, however, by the lack of a practical manufacturing man to put in charge of their operations. The post of superintendent was filled part-time by Mr. Morrison; but he was a busy wholesale druggist, and furthermore, he was not experienced in large-scale chemical making. To these backers of the New York Chemical Manufacturing Company, the arrival in New York of a young chemist fresh from the Sorbonne was most opportune. They promptly offered Martin Kalbfleisch the position of assistant superintendent.

Quite as promptly he accepted their offer. It proved to be a happy combination. The position was indeed a fortunate find for a young foreigner without friends or funds. It put him in close touch with a group of influential citizens, and more than that, it gave him the opportunity to check his theoretical knowledge by the tests of actual plant-scale experience. He worked hard, demonstrated his capabilities, and within a year was put in charge of the enterprise. Under his management the plant was moved to larger quarters way out in the country (now Thirty-second Street, New York City). The output of the original products was greatly increased, and the number of items manufactured more than doubled.

For three years young Kalbfleisch managed the New York Chemical Manufacturing Company. However, he had no intention of remaining the operating man for a group of outside stockholders. He saved his earnings scrupulously, and in 1829 stepped out as a chemical manufacturer in his own right. It is interesting to note that a few years later the merchants who had backed the New York Chemical Manufacturing Company, turned from chemical making to finance and the corporation became the New York Chemical Bank.

In the meantime their former superintendent, Martin Kalbfleisch, had bought some cheap land up in Harlem, near where he lived; built a tiny wooden plant; and engaged in the production of paint pigments. For six years this business steadily, but slowly, grew, until in 1835 when he was tempted by the boom prices being paid for Manhattan Island real estate to sell both his home and his factory and move out to Bridgeport, Connecticut.

Because this change in location carried him too far from his market, it proved to be extremely disadvantageous; but it had one distinct advantage. It started Martin Kalbfleisch upon the manufacture of sulfuric acid, a product that was eventually to become a specialty of his and one that naturally led him into the manufacture of the various sulfates and later on into the production of the other acids, mineral and their salts.

While in Harlem his sense of professional ethics had restrained him from entering into direct competition with his former employers. Out in Connecticut he felt that he had entered a new territory and was accordingly justified in manufacturing the same general line of heavy chemicals. During the five years that he struggled against the geographic handicaps of this location, he installed the apparatus and worked out the processes for the production of these chemicals. In 1840 he moved back into the Metropolitan area, establishing himself in a new and larger plant at Greenpoint, Brooklyn.

The first tenet in Martin Kalbfleisch's chemical creed was that sulfuric acid is the cornerstone of chemical operations. Believing this, he early determined to make the purest, strongest sulfuric acid that might be produced, and he resolved to start only with pure brimstone as a raw material. For many years he endlessly studied the reactions, the equipment, and the controls of this basic chemical process. The result was naturally that Kalbfleisch sulfuric acid soon established a reputation for itself which the company jealously guards to this day.

This deep concern with the strength and purity of his products manifested itself anew as each new item was added to the growing list of Kalbfleisch chemicals. In a short time, therefore, it came to be very well understood throughout the trade that the Kalbfleisch trademark, "Quality First," was not an advertising slogan, but a declaration of business principle.

In 1850 the chemical works of Martin Kalbfleisch moved for the third and last time into what was then the most modern and largest chemical plant in America, between Metropolitan and Grand Avenues, in the town of Bushwick, Long Island. By this time he was mak-

Franklin H. Kalbfleisch

Franklin H. Kalbfleisch's home at Babylon, Long Island, N. Y., where he lived many years, and to which he retired in 1920.

ing sulfuric, muriatic and acetic acids, ammonia, alum, and the various salts of iron, copper, zinc, tin, and nickel. To supply his own great needs for apparatus, he had also established a pottery works, and he had also financially backed the Brookfield Glass Company. His outside business interests expanded with the growth of his chemical enterprises, and he was soon called to serve upon the directorate of a trust company, a bank, and an insurance company.

When he first moved to Greenpoint, he was distressed to find no suitable school for his large family. Characteristically he went to work, organized his neighbors, formed a new school district, rented and repaired an old building, and engaged a competent schoolmaster. Still dissatisfied, he went forward to secure for the district a new, specially built school building.

This interest in public education made an easy introduction into local politics, and he was three times called upon to serve as Supervisor of Bushwick, and later, when that town and Williamsburg were to be joined with Brooklyn, he was one of the commissioners who drew up the charter of consolidation. For six years he was a member of the Brooklyn Board of Aldermen and during half that time served as President. On May 6, 1861, he took office as Mayor of Brooklyn, and on the expiration of this term in 1863, he was sent to Congress, being re-elected to his post in the House of Representatives until 1868 when he once more became Mayor for four consecutive terms. Having in all served as Mayor of Brooklyn for a longer time than any other man, he was nominated by the Republican Party for the Governorship of New York State, but this he refused.

Martin Kalbfleisch left a clear and distinguished impression upon the civic development of Long Island. The time of his greatest political activity was not only a period when corruption was grasping to get its first hold upon municipal affairs; but also when the nation was in the throes of the Civil War. His personal integrity—his forthright honesty won him the nickname of "the Honest Dutchman"—served the cause of honest government long and faithfully; and his high courage was more than once put to the severest test. On one occasion he stood up before an angry, armed mob of draft rioters who had overridden the police, and in plain words called them traitorous cowards, ordering them to disperse. One of the mob's leaders rushed at him. Leaning over, Mayor Kalbfleisch dragged him up on the steps beside him. Silencing the crowd with a gesture, he invited the rioter to tell his story. "Then afterward," he thundered at the mob, "I will tell you your duty as citizens of the Republic!" Ashamed, the draft evader slunk away and his followers quietly scattered.

His refusal of the nomination for Governor was prompted chiefly by his own failing health. For fifty years he had driven himself unsparingly, first in the building up of his chemical business and second in wholehearted public service. In 1869, the firm which for forty years had been known only by his own name became Martin Kalbfleisch & Sons. He retired and the active management devolved upon the three sons who had been associated with him, Albert M., Charles H., and Franklin H. However, he never enjoyed the rest he so richly deserved, for he was an invalid until his death, four years later, in 1873.

The youngest son, Franklin H. Kalbfleisch, born in Bushwick in 1846, had from an early age shown an aptitude for the chemical business and indicated clearly his sense of responsibility to those chemical interests. Educated at the Henry Street Grammar School in New York City and at a boarding school in Warwick, N. Y., he had voluntarily given up college and at the first signs of his father's illness had gone to work in the chemical plant at Bushwick. He started in at the very bottom of the operating units, conscientiously working his way through all the different departments, learning all the products and processes at first hand, and ending in the laboratories. Here he imposed upon himself a stiff course in theoretical chemistry and discovered for himself the value and the meaning of research. Accordingly, when his father was forced to give up his active control, he was well equipped to take over command. The four additional years of Martin Kalbfleisch's invalidism furnished good experience in administration with the opportunity to confer with the man who had founded and built up the business.

With this training and his family tradition behind him, Franklin H. Kalbfleisch had still to prove his own capabilities as a chemical industrialist. He had inherited his father's courage and integrity; but in a number of important respects they were very different personalities. The elder Kalbfleisch, though a strict disciplinarian, was by nature a rare politician, in the finest meaning of that now debased word. Frank Kalbfleisch was a bold and domineering leader, an industrial warrior who thoroughly relished the business battle. He was indeed a rugged individualist, with no more use for trusts and mergers than he had for trade unions. Scrupulously honest in all his dealings, he was without patience with the workman who shirked or the customer who made an unreasonable claim. The former, be he a "hunky" at the wheelbarrow or a crack salesman, he would discharge on the spot. The latter he would simply refuse to sell again.

His concentrated, uncompromising singleness of purpose was shown during the World War when to the despair of his salesmen, he virtually turned his five great chemical plants into factories for the United States Army. When regular customers came clamoring for necessary chemical supplies, he would ask if they did not know the country was at war. In the same vein, he was quixotic in his devotion to a faithful employee or a trusted business friend. Upon his death his family discovered that he was supporting more than a dozen families of ex-Kalbfleisch workmen "even unto the third and fourth generation," and his business associates were often astonished at the liberal terms or special services which he was accustomed to render old customers for what he called "past services rendered the house of Kalbfleisch."

Naturally so doughty a champion broke many a lance in competitive lists. During the Gay Nineties chemical selling was a merry war in which Frank Kalbfleisch neither asked nor gave quarter. The formation of the General Chemical Company reduced the old meleé of competition to what was virtually a three-cornered fight between this merger, Grasselli Chemical Company and Kalbfleisch. This concentrated, rather than mitigated, the ancient habit of warfare.

When, as a young man of twenty-seven, Franklin H. Kalbfleisch succeeded to the post and responsibilities of his father, he plunged into his new duties with an enthusiasm that kept him working about fourteen hours a day. His business responsibilites were not a little complicated by his duties as executor of a large estate of varied interests, and the not unexpected result was that in 1880 he suffered a nervous breakdown. Upon the advice of his physician he sold his New York home, 14 East 55th Street, and moved over to Columbia Heights, Brooklyn. Here it was thought that he would benefit by the better air and, by being between the plant at Bushwick and the offices in Manhattan, would save himself some wear and tear in travelling back and forth. Within two weeks after moving to his new house, two of his children were stricken with diphtheria and died. The shock of this blow to his warm affections broke him up completely, and he was forced to give up active affairs. He promptly sold the ill-omened house on Columbia Heights, and with his wife and other children went South and later to Europe. For five years he was forced to remain away from business.

Then it was he vindicated the family tradition and proved himself a worthy successor of his sire. A man over forty, he opened an office at 19 Liberty Street as a sales agent for chemicals. Within five years he was operating two chemical plants. Only a man of indomitable will and courage to do and to fight for the principles he believed to be right, could have accomplished this rare feat of recovery. Under his leadership the Kalbfleisch business prospered. A program of expansion was successfully executed. By purchase the Erie Chemical Works, the Anatron Chemical Company, and the two plants of Joseph Binns were added as manufacturing units. Steadily and continuously the enterprise grew.

Always the ideal of Martin Kalbfleisch, "Quality First," was the guiding principle. Ceaseless research improved old processes and added new products. Aggressive selling carried the Company into wider and wider fields of service to all sorts of new industries. The World War came as a fitting climax to the long career of Franklin Kalbfleisch, devoted with his notable singleness of purpose to developing one of the great heavy chemical industries of the country.

With wholehearted patriotism he turned these important facilities over to munitions making. Of all their notable war contributions he was perhaps proudest of the work done in gas defense. At the Erie plant sodium manganate was produced for their own use as a purifying reagent in the production of alum, and learning that a British commission had come over seeking a powerful oxidizing agent to use as a gas antidote in masks, the idea was conceived of carrying this operation a step further and producing sodium permanganate. A practical method for the large-scale production of this chemical was devised, and large quantities were furnished to the Government right up to the Armistice.

Amid the hurly-burly of the war activities, he providently brought into the business a younger executive destined to become his successor. He foresaw that the restoration of peace would mean a radical reconstruction of the entire American chemical industry, and he was wise and sincere enough to recognize that the intensive war production was overworking his plant equipment and straining his own physique. Accordingly, after a careful search and very thorough investigation he made a perfectly typical proposal to Harry L. Derby.

"Young man," he said, "I want you to join this organization. You will do whatever it is necessary to do. I don't know what you are worth, so you must set your own salary. If you overvalue yourself, I shall find it out very quickly. If you under-value yourself, I shall think just that much less of you."

Upon these extraordinary terms a bargain was quickly struck which time proved to be a remarkably equitable and mutually advantageous arrangement. For in 1920 the old leader stepped aside for the new. Upon younger, more energetic shoulders were placed the heavy burdens of reconstruction and reorganization.

Mr. Kalbfleisch retired to his beautiful estate at Babylon, a home that he had purchased many years before and where he had lived most happily with his wife, Sarah Pirnie Schenck of New York whom he had married in 1867, and with his two daughters. This home was the sole alternative to his keen interest in business. Here his very simple tastes were fully gratified. He had the happy faculty of complete relaxation, and once away from business he found an absorbing interest in his herd of Jerseys, his greenhouses where he grew hot-house grapes and prize roses, and in later years motoring along the quiet back roads of Long Island. Once he made up his mind to retire, he did so completely; and though he continued to receive daily reports he never interfered with his successor, never made a direct criticism, never offered unsought advice. He marvelled at the vastly expanded chemical horizon of the post-war period, and rejoiced at the growth which his old company made.

ALEXANDER COCHRANE

1802-1865

IN the autumn of 1843, a middle aged Scotch acid maker came to Boston to seek his fortune in the New World. He was a stubby, nubbly Gael with wit and a slow but violent temper. His outstanding characteristic, however, was his super-abundant energy.

In those days, the regular hours in a chemical plant were from seven in the morning till six at night, with a half hour out for lunch; but "overtime" or "after hours," at work or at play, that dynamic little Scotsman was bound to be a hard-driven leader, whipped by his ambitions to hold always the first position. He died sixteen years later, worn out, still a comparatively young man; but in that short time he had gone into business for himself, establishing what was destined to become the most important chemical operation in New England, and training two stalwart sons to carry on the enterprise.

In these days, we should call Alexander Cochrane a "chemical engineer," though he never studied either chemistry or engineering. He had been thoroughly trained, however, in that gruelling school which was personally conducted by British chemical manufacturers of the class of Muspratt, Gossage, Gamble, Deacon, Spence, *et al.* The men of that famous group were real chemical makers. They worked out their own processes themselves, built their own apparatus, turned out their own products, and finally went out and sold them. With such an experience tucked away in his retentive brain and stored up in his dextrous hands, just at a time when the rapid expansion of the New England textile and paper industries was creating a fast-growing demand for heavy chemicals, Alexander Cochrane cannily judged his opportunity in Massachusetts.

First he went to work as a chemist for Lee & Blackburn, but very shortly left to take charge of the chemical operations at the Talbot Chemical Works at Billerica. He was keeping step with local chemical progress, for the old Talbot Dye Works, specialists in logwood, had just branched out into the chemical field with sulfuric and muriatic acids and blue vitriol. It is not difficult to guess how useful this practical Scotch chemical man was in this new chemical plant, nor is it hard to surmise how this experience fired his ambition to employ his talents for his own account. He had learned first-hand how to plan and carry out successfully new methods and processes, and he knew very well that there was a ready market at good prices for the chemicals he could make. Accordingly, he resigned as superintendent of the Talbot chemical plant in 1847 and immediately opened his own works in an ancient stone building at the corner of Medford and Green streets in Malden.

Had there been but a thin streak of superstition in this Scotsman's make-up, he would never have selected this location. His buildings had previously been occupied by a bankrupt dye works and the land had formerly been the old Bellrock cemetery. Several of the old gravestones were still on the property; in fact, one old rectangular tomb of red sandstone, fitted with a wooden cover, was used for a packing counter inside the shed which was the plant's first shipping department. His workmen, however, were not so impervious to eerie influences. The unknown inhabitant of that tomb soon won the reputation of ruling the destiny of the entire operation, and it was considered dangerous to treat him with disrespect. When young Johnny McCarthy, a handy man about the plant, who with reckless hardihood delighted to insult this poor ghost and kick and spit upon his tomb, was severely burned with a solution of hot indigo and sulfuric acid, this superstition became so fixed that Mr. Cochrane had all the headstones and tombs on the property dug up and carted away.

This same Johnny McCarthy is the hero of the only personal anecdote of Alexander Cochrane, Sr., that still lingers in the Merrimac Chemical Company's annals. He was working ten hours for a dollar and a half a day—which itself is a most illuminating commentary on the changes which the past half century have brought to the chemical industry—and not unnaturally he sought a raise. His foreman, James Stewart, said, "Johnny, m'lad, I dinna like t'ask Mister Cochrane for a penny more for ye, me knowin' how worrit he is; but I will na say 'nae' to your askin' y'sel'."

For several weeks young McCarthy pondered this problem, and at last, screwing up his courage, spoke to his employer.

"What," asked Mr. Cochrane, not unkindly, "do you want more money for?"

"For more food to eat and better clothes to wear," came the quick answer, and he showed his work clothes and explained that they wore out so quickly with acid holes that he had very little money left for room and board.

"Food and raiment are necessary evils," commented his employer, his blue eyes sparkling, and turning to the foreman he asked if McCarthy was a good workman.

"Aye," replied Stewart, "he's a guid lad to worrk and tries always to do his verra best."

"Good! Henceforth, give him a dollar seventy five."

The tradition of fair treatment and friendly personal relationship associated with the memory of the company's founder is still cherished. Few chemical corporations have brought up from the ranks so many of their major executives, both in the administrative and operating departments. No company has a brighter roll of long-service employees, son often following his father.

Again, this tradition goes back to Alexander Cochrane himself who, very soon after he embarked in his own business, brought in two sons, Alexander, Jr.

and Hugh, to work with him. Early in the fifties, he took them both into partnership, organizing the firm of A. Cochrane & Company, proprietors of the Malden Chemical Works. When he died on August 11th, 1865, they were ready, trained and in responsible posts, to carry on. Indeed, they had already proved their capabilities, for they had not only shared in the expansion of the business during the Civil War period, but they had also each made a real contribution to the enterprise; Alexander, Jr., in the office and Hugh in the plant.

The two sons were strangely alike, yet they were curiously distinct personalities. Both inherited their father's sandy hair, choleric disposition, and tremendous nervous energy. Both were quick in their movements, and extremely nimble witted. They were both very friendly, sociable men. In very different ways they worked off their excess energy in a number of hobbies. At this point, they differed sharply. Alexander, Jr., was a hunter: Hugh, a yachtsman.

Besides his home on Commonwealth Avenue in Boston, Alexander Cochrane, Jr., owned a farm at Hamilton and a summer home at Pride's Crossing. He was an ardent fisherman and duck hunter, a member of the Aristigooch Salmon Club and Longpoint Company. He was a public-spirited leader in many charities, a vestryman of Trinity Church, and possibly his greatest social service was performed as chairman of the trustees of the Peter Bent Brigham Hospital, a post he held during the critical period when the new hospital was built. His brother Hugh was less active in public affairs, from which his exaggerated dislike of any personal publicity made him shrink, though with scrupulous lack of advertising he was a very generous patron of the arts and loyal supporter of charitable institutions. Hugh was a genial and friendly man, kind-hearted to a degree, who had a host of warm friends and was so greatly beloved by his employees that all the office staff and more than three hundred of the workmen from the plants attended his funeral. He was fond of the sea, and his yacht "Tioga" and schooner "Oenone" were enjoyed especially because of the opportunity for entertaining his friends.

After the death of their father at his summer home at Newburyport on August 11, 1865, these two brothers took up the direction of the business which at the time was known as the Malden Chemical Works, A. Cochrane & Sons, Proprietors, a co-partnership in which they had inherited equal shares. The original product had been sulfuric acid, to which muriatic and nitric acids were soon added. During the fifties, about the time Alexander Cochrane, Sr., took his boys into the firm, the business and equipment of the Newton Chemical Company, situated at Waltham, had been purchased.

This purchase brought to the Malden plant one extremely important piece of apparatus, a platinum still built but a few years previous by an Austrian engineer. This still enabled the Cochranes to produce a higher strength sulfuric acid, using brimstone, than offered by any competitor. It was moved from Waltham to

Malden and twenty years later moved again to Everett where for two decades longer it did active service.

Youth is adventurous and ambitious, and the young Cochranes, after their father's death, determined to venture forth into the dyestuffs field. They selected indigo extract for their initial effort, and after a series of disheartening failures, Hugh, who was already in command at the plant, went abroad to learn from European producers how to make a marketable article. He was away nearly twelve months and after his return was successful in producing a thoroughly satisfactory extract. As indigo was in good demand this new product turned out to be very profitable. It widened their market, increased the customers, and gave them the confidence to undertake research to add other new products.

Munitions demands during the Civil War caused a considerable expansion of the operations at Malden and combined with the growth of the residential section of that city about the plant, made necessary a change in location. For three years, the Cochrane brothers studied this problem, visiting likely situations, inspecting available buildings, warmly debating the pros and cons of every possible location. At last they agreed that the best prospects were offered by an opportunity to buy the New England Chemical Company, competitors in sulfuric acid, and to remove to their plant at Everett. Thus they acquired one large and four small buildings on five and a half acres of marshy flats along the Mystic River. They revamped the chamber apparatus in the large building from a pyrites to a sulfur operation; moved their cherished platinum still and the best of their other equipment; changed their partnership into a corporation under the style of The Cochrane Chemical Company; and during the summer of 1872 established themselves on the site where the vast plant of the Merrimac Division of the Monsanto Chemical Company now stands.

This property is almost at the foreshore of Boston Harbor and at the back boundary of the original purchase run two railways, the main line of the Boston & Maine and the grand junction of the Boston & Albany. The Cochranes had hardly settled down in their new plant before it became plain that very shortly they would be needing more land for expansion. Across the tracks, on slightly higher ground, lay large deposits of clay from which were manufactured high-grade pressed and baked bricks. The fine quality and handsome cherry-red color of these bricks made them high favorites with both builders and architects and many of the finest buildings erected during the past century in and about Boston were constructed of them. Directly opposite the chemical plant was a parcel of land held under a ten-year lease by D. Washburn & Sons. As the Washburns had about exhausted their clay supplies, they were willing to give up their lease, and on March 21, 1874, the chemical company purchased ten and a half acres.

This land had been originally a part of the colonial grant to the Lynde family. The daughter of Seth

Alexander Cochrane, Jr., president, Cochrane Chemical Co., 1872-1917. *From a charcoal drawing by John S. Sargent.*

Robert B. Eaton, founder, Merrimac Chemical Co., 1853-1871.

Lynde was Dr. Sullivan's wife: their daughter had married Alexander Cochrane. Dr. Sullivan had been one of the original Cochrane incorporators.

Later, Seth Lynde was to lend his grandson-in-law, Alexander Cochrane, fifty thousand dollars which so quickly multiplied to a great fortune that, despite his Scotch upbringing and New England surroundings, Alexander Cochrane, Jr., became a sincere devotee of the goddess of Good Luck. His belief in that fickle jade, however, was of the sort summarized by Cromwell when he admonished his soldiers: "Trust in the Lord, and keep your powder dry." Telling the story of this fabulous investment, Alexander Cochrane always emphasized the fortunate chance that opened up this opportunity, rather than his own shrewd foresight.

Shortly after they moved to the Everett plant, the Cochranes began selling acid to a maker of electrical wet batteries named Hubbard. His orders were regular enough, but his payments were extremely irregular. His account finally fell behind as much as several hundred dollars, and Alexander, the business manager, went over to see what might be done about it. This task was made the more difficult by the fact that the Hubbards and their son-in-law were personal friends of his and his wife; and he was finally persuaded, rather against his better business judgment, to accept in payment stock in the little electrical company. Hugh, the plant manager, did not approve of this settlement, so in fairness to him Alexander offered to take over this stock and assume the Hubbard liability on his personal account. He was thus brought into this struggling electrical enterprise. The son-in-law was a professor in Boston University, the technical brains of the business. His name was Alexander Graham Bell.

In this wise, Alexander Cochrane was "on the inside" when the invention of the telephone was perfected and was already a shareholder in the company which became the nucleus of the American Telephone and Telegraph Company. His belief in the future of that invention prompted him to back the initial expansion even to the extent of borrowing funds for this purpose. He thus became one of the largest stockholders in the original Bell Telephone Company, an investment which he attributed to luck, but which made him one of the wealthiest men in New England. In later years, his outside interests weaned him away from his chemical enterprise; but he was long the administrative head of the business and always its guiding financial spirit.

When the Cochrane Chemical Company first took over the ten acres west of the railway tracks, this was a tough looking piece of land with its mud and marshes, its yawning clay pits and dilapidated brick sheds. The attractiveness of the site was in no wise improved by Thompson's slaughterhouse which stood alongside at the waterfront. The aroma, though a stout competitor with hydrogen sulfide fumes, would not have interfered with chemical manufacture but their wastes polluted the water supply so as to menace operations. Accordingly, the Thompsons were bought out with stock in the chemical corporation.

Now the Cochrane Chemical Company was properly set up for expansion. Very rapidly they began to clean up the land across the tracks. Building after building was added: in 1879 the "New South" plant; 1881, the "West Works" and the acetic acid-Glauber's salt plant. This same year the Cambridge Chemical Works was bought and for a time both plants operated. This purchase added another sulfuric acid unit and a new product, ammonia.

In the fall of this year, after vain efforts to get well trained chemists with American plant experience, Augustus Olsen and John Enequist, were brought over from Sweden as Superintendents. C. R. Gyzander was also engaged as chemist in charge of the laboratory. In 1882, Frank G. Stantial became superintendent at Malden and Cambridge, and three years later was given James Lund as assistant. When Olsen left in 1888, Stantial moved to Everett and Lund succeeded him in charge of the two smaller operations.

Growth of the operations and birth of additional products is told in the successive building of plants at Everett. In 1892, the manufacture of indigo extract was moved to a new building and the old Malden plant definitely abandoned. A sulfate of alumina plant was built in 1893 (re-built 1911); ammonia, 1894; a new acetic acid building, 1895; No. 18 Building, 1897; sodium and sulfide, 1903; a contact acid plant in 1907; and glue works in 1915.

In the spring of 1902, Hugh Cochrane, who throughout the past thirty years had directed this sustained program of expansion, returned from Europe a seriously ill man. During the summer he recovered and in the fall seemed to regain his accustomed vigor; but early in the winter contracted influenza and died at his home in Boston, January, 1904.

It now fell upon one of Alexander Cochrane's sons-in-law, Lindsley Loring, actively to represent the family interests. He became vice-president and later treasurer. Intelligently and energetically he took up the duties of the active administration of a large and growing chemical manufacturing business. Since the death of his father, Alexander Cochrane had served the company as president. His brother had left no sons and none of his own three boys was chemically inclined. He recognized clearly that the World War was not only presenting golden opportunities for aggressive development, but that it was also entirely remaking our chemical industry. He felt too old to buckle on armour. The future of the business demanded a reorganization. Accordingly, in 1917, the Cochrane Chemical Company was merged with the Merrimac Chemical Company.

It is now necessary to go back again to the Talbot Chemical Works, where Alexander Cochrane, Sr., made his start in the American chemical industry, and even to their predecessor and one-time competitor, the Woburn Chemical Works. The year is 1853. The growth of the New England textile and paper industries, which fifteen years later created the market for Cochrane's chemical operations, had even then reached a point where a rival to the Talbots (established in 1840) might reasonably be expected to prosper. At least, so argued Robert B. Eaton, an importer with first-hand knowledge of the chemical needs of these manufacturing consumers.

Mr. Eaton had, however, practically no knowledge of chemical making; but this did not dismay him. With a staff consisting of a good plumber and a couple of experienced chemical workmen he went to work to produce sulfuric acid in a tiny plant at Woburn.

How he did it nobody now knows. He not only made sulfuric acid and sold it profitably; but he also in time produced muriatic acid, soda ash and salt cake, Glauber's salt and nitric acid, and tin crystals. The initial output was small and the earliest operations must have been unbelievably primitive; but Robert Eaton was ingenious and persevering. His processes improved and his production grew. As he was thoroughly immersed in his manufacturing problems, he turned over his sales to his good friends Candler, Foster & Co.

At the outbreak of the Civil War, an opportunity for timely expansion presented itself. The sulfuric acid production of the Merrimac Manufacturing Company of Lowell, and Mr. Eaton's Woburn works were all combined on November 27, 1863, under the corporate name of the Merrimac Chemical Company. John W. Chandler, the former sales agent, was the first president; the treasurer was C. D. Kellogg; Robert Eaton and W. L. Candler were incorporators and directors. A little later, Page Eaton and Charles O. Foster joined the board. Mr. Eaton continued at the plant and Messrs. Chandler and Foster at the offices in Boston.

After the Civil War, industrial New England blossomed in a heyday of prosperity. Additions to the productive capacity of the chemical plant at Woburn required additional capital, so the stock shares were increased; and in 1871, the Howards became financially interested in the business. C. T. Howard was elected treasurer and A. P. Howard, a director. At the same time, Dr. Charles McBurney was added to the board. In 1877, Barthold Schlesinger, a former partner of Naylor and Company, of New York, and already a large stockholder in Merrimac, became a director. In 1884, he became president and held office till he died in 1902.

Throughout this period the active management was in the hands of the Howard Brothers, A. P. Howard directing the production and C. T. Howard in charge of the office. In 1889, Henry Howard joined the staff as chemist and upon the death of A. P. Howard, his father, succeeded him as superintendent.

During the Howard regime, Merrimac materially improved its manufacturing operations and considerably diversified its products. Theirs was a chemically-minded administration, primarily concerned with production; and they laid down very solid foundations under the plants. Alum and sulfate of alumina were produced in 1886; bisulfite of soda and iron nitrate in 1887; aluminum chloride (the first made in America) in 1888. That same year a pyrites process, developed in their own works, was put into successful production of sulfuric acid. Silicate of soda was made in 1890.

Other processes and products were worked out and acquired. In 1896, having bought American rights to the Bayer process, aluminum hydrate was put on the market. Three years later, Merrimac, through the purchase of the works and business of William H. Swift & Company, East Boston, became manufacturers of acetic acid, insecticides, and dry colors. From a Russian company, the rights to the Tentelew System were acquired, and a contact unit for the production of sulfuric acid was erected and put into operation in 1907. During the World War, in alliance with the New England Gas & Coke Company, Merrimac erected a plant for the manufacture of phenol and the high explosives, picric acid and trinitrotoluol. Operations on a large scale were continued until the close of the war.

Merrimac's destinies during the World War were in the hands of a farsighted, quiet-spoken, capable New Englander of faultless chemical heritage. 'Way back in 1787, through the General Court of Massachusetts, Caleb Wilder and his associate William Forbisher had patriotically made available to their fellow citizens certain improvements which they had perfected in the processes of what was New England's first important chemical industry, the production of pot-ashes. Caleb's great grandson, Salmon W. Wilder, having graduated with the very first class in chemical engineering at the Massachusetts Institute of Technology in 1891, had joined the Merrimac staff in 1897. In 1899, he had been made manager; in 1903, treasurer; in 1906, president, a post he filled till 1928 when he became chairman.

At the time of his death, Mr. Schlesinger was the largest single shareholder in the company, and Mr. Wilder with William A. Russell, of The Russell Company of Boston, bought this stock and with it control. It was in the midst of bustling war activities, under Mr. Wilder's leadership, in 1917, that the Merrimac Chemical Company took over the Cochrane Chemical Company.

Though technically a purchase, this transaction became in reality a fusing of the two leading chemical companies of New England. The personnel of the consolidated operation reveals how intimate was this amalgamation. Mr. Wilder remained president, but the two chief plant executives of Cochrane, Frank G. Stantial and James Lund, became vice-presidents, and the Cochrane treasurer, Mr. Loring, served Merrimac in the same capacity.

There is much that is distinctively of New England in the birth, the growth, the merger of these two old New England chemical companies. Their makers were New Englanders or true adopted sons of New England, men connected with old, prominent Bay Colony families with traditionally important financial and industrial connections. Each company was initiated in direct response to the demand of New England industries for chemicals, and their development was predicated upon the widening use of chemicals within a limited area. In this field they gained a dominating supremacy. This they held against vigorous competitive invasions of

C. Robert Gyzander

William J. Webber

James Lund

Frank G. Stantial

Four Cochrane-Merrimac Executives

Charles T. Howard

Alonzo P. Howard

Barthold Schlesinger

Salmon W. Wilder

Four Executives of the Merrimac Chemical Company.

their territory. Both companies never failed year in and out, throughout booms and depressions, to earn a dividend. This heritance has been carried on since in 1926 Merrimac merged with the Monsanto Chemical Company; and there are customers on the books today whose active accounts go back through three generations while in both plant and office there are grandsons of men who worked with the Cochranes and the Howards.

JAMES JAY MAPES

1805-1866

NO other pioneer of the American chemical industry lived life with the zest and relish of James J. Mapes. That lusty vigor of mind and body which Plutarch noted as one of the prerequisites of great accomplishment, catapulted him through a dozen diverse chemical careers, so that he was not only the first maker in this country of both superphosphate and complete mixed fertilizers, but also the first chemical engineer to open a consulting office in New York, and the first American manufacturer of a synthetic tanning agent and of Epsom salts from the hydrobisilicate of magnesium. He was moreover, the inventor of important chemical processes and apparatus used in sugar refining, in color making, dyeing, distilling, and steel tempering. He introduced practical methods of chemical control into the then-infant industries of daguerreotyping and electrotyping. An analyst of skilled resourcefulness he perfected methods for beer and wine determinations which were for many years standard practice. He was one of our first, and is still remembered as one of our greatest, expert court witnesses. For many years he taught both inorganic and organic chemistry. He originated, and upon the basis of his own practical experience organized, the first courses in agricultural chemistry.

Unsatisfied by activities that would have completely consumed the energies of three or four active men, he was for two years editor of *The American Repository of Arts, Sciences and Manufactures*, resigning only to found and edit a magazine of his own, *The Working Farmer*. He bought an outworn farm and transformed it into what became literally the model of the modern Agricultural Experiment Station. He invented an improved sub-soil plough and engaged in the nursery business. He served as the very active officer of a great number of scientific associations. He organized several night schools, forerunners of the Cooper Institute, and was for years a pillar of the society that grew into the American Museum of Natural History. He wrote articles for practically every scientific journal and farm paper of his day and made more than a hundred and fifty public addresses. As a youth of seventeen he delivered a course of lectures on military tactics illustrated by a relief map with moving figures depicting Napoleon's disastrous Moscow campaign, and fifty years later Lincoln called him to Washington to confer with the Union generals on their plans to encompass Richmond.

Not content with all his scientific, agricultural, and industrial pursuits, as a hobby he painted in oils on canvas and miniatures in water colors, and two of his portraits were hung at exhibitions of the National Academy. He was a first class amateur musician and he wrote jolly doggerel verse for the amusement of his friends and family.

Professor Mapes was a famous wit with rare gifts of mimicry and dialect, and in days when dining was a fine art, he was known as an epicure. He dearly loved good company, entertained constantly, and was always in demand as a dinner companion. He was a member of the old New York Club, the St. Nicholas Society, the Sketch Club and others; and among his close friends were Washington Irving, William Cullen Bryant, Horace Greeley, Richard W. Gilder, and Marion Harland from the world of letters; Samuel F. B. Morse, inventor of the telegraph, Frank Forester, the sportsman, John Ericsson, the engineer who built the "Monitor," Professor Tenny of Williams, and from the world of affairs John L. Morton, Charles Ingham, Thomas J. Cummings, Henry Inman, A. B. Durand, and his own cousin, Samuel Waldo.

Till the day of his death James J. Mapes lived with true Elizabethan gusto. Even in the expansive epoch of our history during which he lived he seems quite un-American, displaying rather the versatile talents and catholic tastes typical of the man of the Italian Renaissance. He lived well and was even lavish in his hospitality; but his comfortable income came chiefly from his personal services as chemist and consultant and expert witness, as author and editor and lecturer, rather

James Jay Mapes

What Prof. Mapes Did in 1847 The Good Farmer Does in 1927

"I'll go to the crop", said Prof. James J. Mapes, originator of Mapes Manures. "I'll ask the crop to tell me what fertilizer it needs". Although famous as an expert chemist, Prof. Mapes knew that he could not depend on chemical analysis alone. So, in 1847, he bought a farm to check up, in the field, what he had learned in the laboratory.

The good farmer of today knows, as Prof. Mapes did eighty years ago, that the crop is the best judge of fertilizer values. He knows that two fertilizers of the same analysis may give widely different results because of the different materials from which they are formulated. So he buys his fertilizer on the basis of crop results, not on analysis alone.

That is why Mapes users are so loyal to Mapes Manures. And that is why more good farmers every year are becoming Mapes users. We go to the crop; we ask it what materials it likes best; we put these materials into Mapes Manures. Mapes Manures are made to grow good crops—not to sell at a price. They are first made right, then priced as low as possible.

If you are not a Mapes user, try Mapes this year. Compare the crop *yield*; compare the crop *quality*; compare the crop *profits* with the result from any other fertilizer you can buy. Mapes "costs little more — worth much more".

Just Mail This Coupon Today

Write today for a list of the crop brands and prices of Mapes Manures. You'll be surprised at the little difference in cost between Mapes brands and other brands. Mapes "costs little more—worth much more". Please tell us what crops you plan to fertilize so that we can be of the greatest possible service to you in selecting the right brand to suit your special needs.

The Mapes Formula and Peruvian Guano Co., Dept. 13,
270 Madison Ave., New York, N.Y.

Without obligating me in any way, please send me your list of crop brands and prices.

I use..........tons of fertilizer on the following crops: ....

....

My name is..

P. O... *State*......

MAPES
Manures

than from his business ventures either as a merchant or a manufacturer. A lesser man would have been dissipated by his multitudinous interests; but so tremendous were his energies and so great his mental resources that James J. Mapes was able during his comparatively short life to carve his name in bold letters within half a dozen niches in our Chemical Hall of Fame.

The Mapes family from whence sprang this original, universal genius was probably of Welsh origin. The Doomsday Book records that one branch owned extensive lands in Herefordshire before the Norman conquest, and an early, distinguished member of the line was Sir Walter de Mapes, scholar, courtier, Archbishop of Oxford, who as court poet collected the popular stories of King Arthur and his Knights of the Round Table, weaving together the legends of their quest for the Holy Grail and creating the character of Sir Galahad.

The first of this ancient family to come to America was John Mapes who landed in 1634 but died without issue so that the American branch is descended from Thomas, one of the seven men who in 1640 under the Rev. John Youngs from New Haven made the first settlement on Long Island at Southold. Here on Thomas Mapes' tract of over three thousand acres the American branch took root and settled down as farmers. They prospered and mutiplied, and twenty men of the family fought in the Revolutionary War.

Professor Mapes' father, Jonas Mapes, was born on

Long Island in 1768 and here he was also born, at Maspeth on May 29, 1806. During the War of 1812 his father as major general was in command of all the defense troops in New York area. The family removed to the city, and after the war General Mapes entered business as an importer and merchant tailor. He was shortly elected an alderman, was appointed treasurer of the committee which received General Lafayette in 1824, and accompanied him by his special invitation on his triumphal tour of the country. He was a founder and for many years a director of the Bank of Savings (N.Y.), in Bleecker Street and was intimately associated with DeWitt Clinton in the Erie Canal project which was destined to establish firmly the commercial supremacy of New York over its seaboard rivals. Thus the family occupied an important position in the life of the city and the young James was brought up under the most favorable social conditions.

At the tender age of eight the precocious boy showed plainly the bent of his interest towards chemistry when, after being taken to a lecture on the subject, he made illuminating gas in a retort improvised out of a piece of clay pipe. At eleven he was sent down to Hempstead, Long Island, to board with the family of the illustrious English reformer, William Cobbett, and to attend the famous classical school of Dr. Timothy Clowes. This experience left lasting impressions upon the youthful Mapes, for though he was in the main self-educated, particularly in chemistry, nevertheless he owed to Cobbett his life-long interest in agriculture and to Clowes his taste for belles-lettres.

While still in his teens he entered the mercantile house of his father; but in 1827 General Mapes died, and the son having attained his majority launched out in business for himself as a cloth merchant. That same year he married—October 10, 1827—Sophia Furman, daughter of the late Judge Garrett Furman, a distinguished member of another old, prominent Long Island family.

James Mapes continued in the mercantile business for the next seven years; but all during this time he was studying and experimenting in chemistry. In 1832 he invented several notable improvements in the methods of refining sugar and subsequently he perfected apparatus for the more efficient extraction of sugar both from cane and from molasses. Having won a substantial reputation as a chemist, he forsook business in 1834 and for the next twelve years supported himself by his consulting and analytical work.

One of his first clients was the New York State Senate which commissioned him to make a series of analyses of beers. During this period of his career he manufactured Epsom salts and produced from this material a chemical tanning agent designed to replace natural hemlock extract. He made a number of important contributions to a wide variety of chemical industrial processes and helped Seth Boyden in the experiments which led to his invention of malleable iron. As an amateur artist he became much interested in colors and produced several new pigments of exceptional brilliancy and permanence which had never before been available in America. His most remunerative work, however, was as an expert witness in chemical patent litigation. His sound knowledge and wide practical experience enabled him tellingly to marshall and present his facts. His alert intelligence and unfailing composure withstood the most gruelling cross-examination. His good humored wit was a weapon that the ablest lawyers of his day learned ruefully to respect.

Along with his chemical practice, James Mapes served as Professor of Chemistry and Natural Philosophy of the National Academy of Design and later held a similar position in the American Institute of which he was for many years the vice president. In 1844 he was elected president of the Mechanics Institute, and through these organizations he zealously promoted night schools for adult education. He was interested also in natural history and was elected a permanent member of the New York Lyceum and of the National Institute of Washington.

Gradually, however, Professor Mapes' chief interest began to focus more and more upon scientific agriculture. His studies convinced him that "our farming methods are primitive even to the verge of barbarism." Accordingly, he embarked upon a crusade to introduce sub-soil drainage, crop rotation, and the use of chemical fertilizers. The good dirt farmers of his day greeted his theories with hoots and cat-calls, so to convince them that there was something practical to be learned

Charles Victor Mapes

Charles Halsted Mapes

from "book farming," he purchased a wornout farm in the Weequahic section of Newark, New Jersey and set to work to prove theory by practice. This was in 1847, and from thenceforward his life was devoted to this cause.

He took up this herculean labor of overthrowing long established farming habits and conquering farmers' inertia with typical enthusiasm and energy. He established a magazine, *The Working Farmer*. He organized state and county agricultural societies. He went all over the eastern states talking at meetings of farmers. He was the first to advocate a national Department of Agriculture with a Secretary who should be a cabinet officer.

But most important of all, he transformed the poor wornout soil of his Newark farm into highly productive acres. He grew both vegetables and field crops under carefully controlled conditions. He selected plants and trees for vigor and yield. He sold seed and nursery stock. He took pupils in small groups of three or four for a month's "short course" in scientific farming. He became the pioneer American manufacturer of chemical fertilizers.

His bold theories had been convincingly confirmed by the work of Baron von Liebig; and since he was familiar with these experiments, he naturally began to make superphosphate, as suggested by the great German agricultural chemist, by treating bones with sulfuric acid. As early as 1849 he produced this superphosphate for use on his own land. Three years later he prepared an "improved superphosphate of lime" from the charred bone which was a waste product from the sugar refineries to which he added sulfate of ammonia and Peruvian guano, which also had been treated with acid to prevent the loss of volatile ammonia and make the phosphate more immediately available for plant food. This was undoubtedly the first mixed chemical fertilizer made commercially in America, but his patent applications were held up and not issued until 1859. It is quite typical of Professor Mapes that once his patent rights were assured he never endeavored to enforce them, but on the contrary encouraged others to make and market similar fertilizer mixtures.

Till the day of his death, for he literally died in harness on January 10, 1866 at his New York home, he worked manfully and to good purpose to further the cause of better farming. For several years his health had been failing and he was never one to save his energies. The task he put upon his physique was increased by his excessive weight, and in the end his heart gave way under the strain. His death was undoubtedly hastened by his exertions when, as an expert witness in an important patent suit in Boston, it was necessary several times a day for him to climb a long, steep flight of stairs. He came up weak and panting, but his good humor could not be downed, and he exclaimed: "What a pity that when the United States owns so much land they are obliged to hold court so high up in the air."

James J. Mapes was survived by his wife, three daughters, and one son; and the vigor and versatility of this great man has carried on through three generations of a family that has produced an exceptional percentage of distinguished individuals. His favorite child was a daughter, Mary Mapes Dodge, the author of the children's classic, "Hans Brinker," and for thirty years editor of *St. Nicholas,* whose keen mind and rare social and literary gifts owed much indeed to her father's influence, training, and encouragement. His son, as we shall see, succeeded him as a fertilizer manufacturer and won in his own right high honors as an agricultural chemist. Among his grandchildren are found a distinguished physician, James Jay Mapes II, who was the first to use diphtheria antitoxin in America and who made his name well known even in a very brief career, Victor Mapes, the playwright, and Clive Mapes who left the family business to become a pioneer in the radio industry.

The father of this distinguished trio, Charles Victor Mapes, inherited from his father a bent for natural science and an abiding interest in agricultural problems. He had the same versatile, brilliant mind, the same artistic and musical tastes. He may have determined to concentrate his efforts or he may have lacked his father's insatiable chemical curiosity; but at all events he centered his interests upon the fertilizer business with the result that he made a conspicuous success of this enterprise.

Charles Victor Mapes was born in New York, July 4, 1836, and the first ten years of his boyhood were spent in the city after which the family moved to the farm in New Jersey. Here he fitted up a little chemical laboratory in his bedroom and grew up a born chemist in the midst of intensive agricultural experiments. He went to Harvard and at one time planned to study medicine, but in 1858 entered the counting room of a firm of wholesale grocers in New York. The next year he established himself in Newark and with one of his employers as the silent partner who furnished the working capital, began the manufacture of farm implements and fertilizers. He took over the publication of his father's paper, *The Working Farmer;* and settled down seriously to what was to be his life work.

In 1877 the Mapes Formula and Peruvian Guano Company was organized and he became vice president and general manager. Offices were maintained in New York and the factory was located in Newark. When the New York Chemical and Fertilizer Exchange was organized he was elected the first president. He worked long and faithfully to provide an open market for these materials, an effort that failed eventually because manufactured wares do not lend themselves economically to exchange trading in future deliveries. Like his father he often spoke at meetings of farmers and wrote articles on practical farming problems.

But Charles Victor Mapes' great chemical contribution was the special fertilizer particularly formulated to supply the deficiencies of certain types of soils or to provide the special plant food requirements of certain crops. In 1874 he prepared a fertilizer for Irish potatoes and marketed it to southern truck growers. This was the first special-crop mixture ever made in the United States and was followed by other specialized fertilizers for tobacco, citrus fruits, grapes, cotton, corn, etc. Much of the data upon which these various formulas were based came from experiments which Mr. Mapes made in collaboration with Professor W. O. Atwater. These were the first tests carried on in America under anything approaching modern methods of field fertilizer experiments and embraced careful checks on various natural manures, fertilizer ingredients, and complete mixed fertilizers.

So the tradition of scientific agriculture established by the father was pushed forward by the son, and with all due allowance for the advertising point of view, the following quotation from a booklet on citrus tree fertilizers issued in 1914 by the grandson, Charles Halsted Mapes, is interesting as confirming the family doctrine:

"We have been manufacturing fertilizers for over fifty years and have been furnishing them in Florida now about forty-five years, but there are so many newcomers in the state that it has occurred to us that they might be interested in hearing a little in detail what our company has done for fruit fertilizing, and our reputation in the state. The few fertilizers that were selling when we entered the Florida business have entirely disappeared from view. From the start, we established the business on a thoroughly scientific basis; in fact, we were the pioneers of scientific fertilizing in Florida. There is not an improvement in the business which we were not the first to suggest and carry out.

"We were the first to advocate the use of Sulfate of Potash

and forms free from Muriates and Chlorides, so deleterious in saccharine and starch formation, long before the stations made any difference in valuations between the Sulfate of Potash and the cheaper Muriates of Potash.

"We were first to recommend the two distinct types of fertilizer, our Orange Tree Manure for wood growth, for growing the tree, and the Fruit and Vine Manure to develop the tree and fruiting power. This system proved itself such an absolute success that it was soon adopted by all other manufacturers.

"We were first to practise and call attention to the great value of varied as well as choice forms of plant food, and the seasoning process which takes place on goods made up and stored in piles, before bagging, long in advance of their use, so that they are thoroughly mellowed and ready to give immediate and best results on application to the crop.

"We were, and are, practically alone in using the choicest phosphatic materials known, Bone and Guano, as a basis for our goods and thereby avoiding acidity of acid phosphate rock goods."

Likewise, it is interesting to compare the formulas and recommendations for the use of Mapes Fruit and Vine Manure with what is practiced today. The Mapes formula had a guaranteed analysis of ammonia, 2-5%; available phosphoric acid, 5-7%; potash, 10-12%, and the application of three pounds for the first year, six pounds for the second, and ten pounds for the third was suggested. Van Slyke recommends: 500 pounds of sodium nitrate, 500 of superphosphate, and 125 of potassium sulfate with applications of three pounds the first year, four and a half pounds the second, and six pounds the third. The principals remain much the same; but the total amount of actual plant food had been virtually trebled and the proportions of nitrogen and potash considerably increased.

Charles Victor Mapes married Martha Meeker Halsted on June 25, 1863 and they had three sons. Two of these, Charles Halsted Mapes and Clive Spencer Mapes went with their father into the fertilizer company. The first became active manager and served till the business was finally wound up in 1926, and the direct connection of their remarkably gifted family with the chemical industry was thus terminated nearly one hundred years after James Jay Mapes hung out his shingle as the first consulting chemical engineer ever to open offices in New York.

EUGENE RAMIRO GRASSELLI

1810-1882

IN 1839, a chemical manufactory had been established and the production of sulfuric acid, alum, and Le Blanc soda began at Cincinnati. That simple statement of historic fact reveals a man of clear-sighted chemical vision and uncommon courage. His vision, recognizing that capable chemical manufacturers were established in the industrial centers of the Atlantic seaboard, saw beyond the heavily wooded slopes of the Alleghany Mountains—a barrier then crossed only in creaking ox carts—and foresaw correctly the economic empire of the Middle West. His courage was undaunted by the peculiar risks and difficulties of chemical manufacturing on the frontier far from the base of all industrial supplies.

That man of vision and courage was Eugene Ramiro Grasselli. But he was no reckless dreamer. His plans for a chemical-making enterprise at that outpost on the Ohio River had been carefully thought through to a definite economic conclusion based upon sound chemical logic.

Eugene Ramiro Grasselli

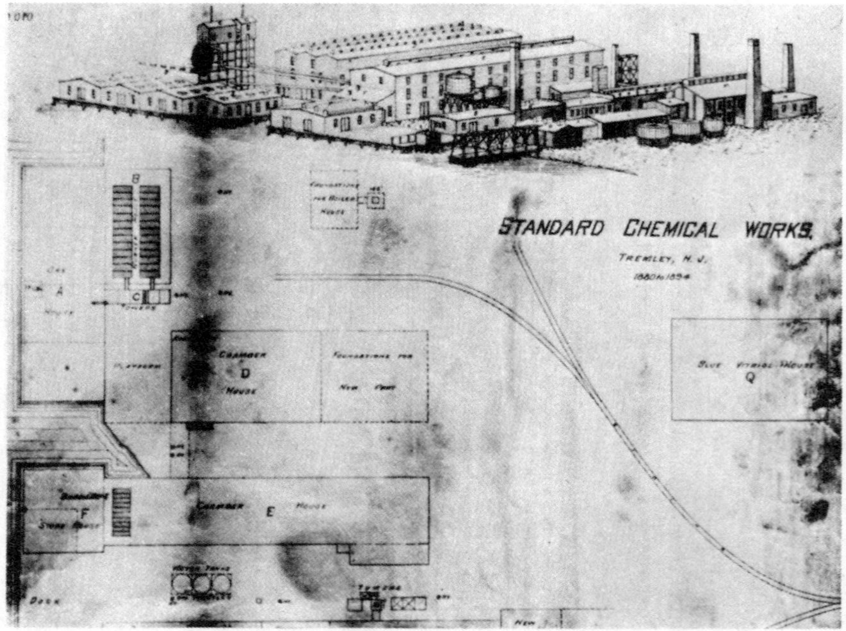

Plan of the Standard Chemical Works, 1880 to 1894, foundation upon which were built the big scale operations at Grasselli, N. J.

For he was a trained chemist—educated at the Universities of Strasburg and Heidelberg—the son of a long line of chemical manufacturers, with practical experience gained in a gruelling apprenticeship served under his own father's direction in the family plant.

The story of Eugene Ramiro Grasselli's success is not cut from the usual American patterns. He was neither a sturdy son of the Pilgrim Fathers nor a scion of Virginia's first families. He was not born in a log cabin. His boyhood was not spent barefoot on the farm. He was not even a friendless, penniless immigrant seeking his fortune in this land of opportunities. He came of an Italian family which, since medieval times, had been druggists and chemists. The ancestral records go back to 1440 when at Torno, on Lake Como, the Grassellis were established as makers of medicine and perfumeries, chemicals and gunpowder.

At Torno his father, Giovanni Angelo Grasselli, was born in 1781; but as a young man he moved to Strasburg, Alsace, France, there to launch himself independently in the chemical business. Unsettled conditions in Northern Italy and a young man's desire to win his own success, doubtless prompted this migration. Soon after the plant at Strasburg had been established, in order to avoid the prohibitive import duties that Germany then levied against France, a plant was opened in 1810 at Wohlgelegen near Mannheim, Germany. In both plants sulfuric acid was the principal product, and the history of the Verein Chemische Fabrik, Mannheim, gives credit to Giovanni Angelo Grasselli for having been the first to bring Sicilian brimstone into Germany. Muriatic acid was also produced and the common salts of both these acids.

It was in Strasburg, January 31, 1810, that Eugene Ramiro Grasselli was born. There he was raised under French influence, and when he came to America, although he had attended German universities, French was his "native" tongue. He landed in Philadelphia in 1837 and found employment there with Farr and Kunzi, remaining with them two years. Doubtless from the first he considered the connection but temporary, giving him an opportunity to become acclimated to the strange land and to study the American chemical field. He came to this country inspired by the ambition, as his father had been when he left the family's ancient headquarters on Lake Como, to supply chemicals of his own manufacture to a new and growing market.

Accordingly, as early in the spring of 1839 as it was possible to travel, young Grasselli left Philadelphia. He took the train as far west as it then went, to Harrisburg; on to the foot-hills of the Alleghanies by canal; across the mountains by ox cart, and so into Pittsburgh. From Pittsburgh transportation on the Ohio River was available by river barge, and he took this opportunity to study locations which came under his observation on this long journey of some five hundred miles to Cincinnati.

Of the various locations observed he found Cincinnati, a community of 42,000, already a thriving manufacturing center and decided to settle there. Diversified industrial development had taken place, and cattle, which at this time grazed in great numbers on the open prairies of Ohio, were brought to Cincinnati and the packing industry there gave rise to the manufacture of its by-products into soap and candles. Here was an immediate chemical market, while for the future the opening of the territory to the west and south held forth a promising prospect.

A few hundred feet from the city limits of Cincinnati he found the building site he was seeking, a triangular piece of land located on the Miami and Erie Canal which drew its waters from Lake Erie at Toledo, continued its course to Cincinnati, and emptied into the Ohio River. Projected railroad facilities were provided by a charter of the Little Miami Railroad granted in 1836, the first section of which was opened in 1843.

On this site he established his first factory in 1839. The office and factory building faced east and west, the factory facing west 75 feet frontage; facing east 90 feet; along Martin Street, 225 feet; along East Front Street, 345 feet. Sulfuric acid chambers covered a lot 165 feet frontage on the north side of Martin Street and extended back 30 feet to a hillside. The construction was a combination of stone, brick, and wood.

Sulfuric acid was the key product, used at first chiefly in further chemical processes, while alum, soda ash, and Glauber's salt were the principal items sold. Later nitric and muriatic acids and ammonia were produced, together with a number of pharmaceutical preparations. The market for these products expanded rapidly and the bold venture prospered. Within six years a direct competitor appeared in Cincinnati, The Marsh & Harwood Chemical Company, established by David M. Marsh and Edward Harwood, destined years later to become allies, even associates, of the Grasselli interests.

In 1845, however, Eugene Grasselli's immediate answer to this competition was to improve his own processes. From France he imported the first platinum acid refining still to be erected in the Middle West. He personally laid the brick and adjusted the loam cushion upon which this valuable piece of apparatus was to rest. A few years later, in apparatus of his own design and making, he began the manufacture of chloroform.

This item was to assume great importance during the Civil War, which created not only a great opportunity for the sale of chemicals but also undreamed of difficulties in their manufacture. Since the lower Mississippi was in the hands of the Confederates, Sicilian brimstone had to be imported through Philadelphia. So added to the extraordinary rail freight charges across the mountains was a war tax of $6 per ton. Chile nitrate, commandeered for munitions, was virtually unobtainable. Currency was deflated and prices fluctuated widely. Shipments both to and from Cincinnati became highly uncertain. In the face of these difficulties, Eugene Grasselli determined to forge ahead.

During the war period Grasselli sales, formerly handled in the West by Allen & Company, wholesale druggists, and by James P. Morgan & Company in the East, had been taken over directly by Grasselli himself. An efficient sales organization was developed by R. H. Andrews a shrewd merchant and competent organizer. At the close of the war a rapid industrial expansion began in the Middle West. There were developments that in particular opened up new, great consuming fields for sulfuric acid. Petroleum refining, steel treating (especially the cleaning of wire and nails), and the manufacture of ammonium sulfate from the ammonia in the wash water of the gas works, were all in their infancy but growing rapidly. New competition began to appear. At Pittsburgh James Irwin, an Ohio River steamboat captain, was erecting a new acid plant. A group of Cleveland petroleum refiners, including Hussey & McBride, W. C. Schofield, and W. P. Eells, president of the Commercial Bank, had incorporated the Cleveland Chemical Works.

Eugene Grasselli determined to meet these developments aggressively. In 1865 he built a new plant at Cleveland, Ohio, and since the location was obviously better, both in respect to raw materials and to customers, he resolved to move his headquarters there.

This meant uprooting his family. He had married Fredericka Eisenbarth in Philadelphia and they had eight children. The eldest, a daughter, Lucretia, had married Daniel Bailey, a promising young mechanical engineer who was now connected with the business.

The fifth child, a son, Caesar Augustin Grasselli, then a boy of seventeen, was already working with his father; in fact, at fifteen he left school to go to work in the plant. Many years later when this same boy became the Chairman of the Board of the great Grasselli Chemical Company, he wrote: "I cannot remember the time when I was not interested in chemistry and did not expect to follow my father in this business."

Father and son had had a long serious talk. Eugene Grasselli pointed out to his son that a thoroughgoing apprenticeship through every operating and administrative department would give him, if supplemented by some formal instruction in chemistry, a training of great practical value. With that charm and sincerity for which he was always noted he pictured to the youthful Caesar the romance of the chemical industry, its great service to civilization as the handmaiden of all manufacturing activity. This was rich fuel to fire the enthusiasm of a boy who, by long inheritance and strong inclination, was foreordained to a chemical career.

The next day young C. A. Grasselli donned overalls and went to work in the acid plant. Nights he studied chemistry by special arrangement with a professor from the Karlsruhe University. Always he was under the friendly, watchful eye of a father who instructed him in practical mechanics and engineering in the machine shop, made opportunity for new experiments, disclosed process workings, and finally taught him the commercial end of the business through office training, instructing him in new duties and putting on him always increasing responsibilities. Thus, as has so frequently happened in the American chemical industry, the enterprise founded by the father as a personal business was carried forward by his son to become a great corporation.

C. A. Grasselli's first opportunity came with the building of the new Cleveland Plant. Land had been purchased on the Cuyahoga River. The plans were drawn by Eugene Grasselli himself, and here he again took advantage of a location beside a river and beneath a high hill. The execution of these plans he put in the hands of his engineer son-in-law, Daniel Bailey. The boy, Caesar, went along as his assistant and on that construction job worked as a brick-layer and stone mason, plumber, pipe-fitter and a tin-smith, mechanic and boilerman. Literally he knew that plant from the ground up.

Eugene Grasselli was actively engaged in the operation of the business in Cincinnati. Daniel Bailey supervised the building of the new plant on Independence Road in Cleveland. In testimony of the exacting skill of the engineering and soundness of the plans from which the plant was built, the buildings erected during this original construction still stand and are useful today in the Cleveland Works. They stand as a tribute to the efficiency and the painstaking, conscientious execution of the work entrusted to Daniel Bailey. Through his long life and connection with these interests he was one of the main supporters of Caesar Grasselli in his many activities.

By the spring of 1867 the new plant was ready to

go into production. The Grasselli family moved to Cleveland, Daniel Bailey returning to Cincinnati to take up his duties at that point. From an operating point of view the beginning was auspicious. Eugene Grasselli had made the most of his wide, practical experience, nor had he hesitated to introduce innovations. Up to that time sulfuric acid chambers in this country had been soldered, but he had brought to Cleveland a Frenchman named Valiant, skilled in the new art of burning overhand lead seams. Output in the new apparatus exceeded even calculated capacities. But the booming chemical demand which had prompted the new plant collapsed suddenly in the post-war panic of 1867. After six months' operation, in the first week of January, 1868, the total sales were sixty-eight cents and the cash receipts seventy-five cents.

Their troubles were aggravated by an epizootic epidemic, a sort of equine influenza that paralyzed the horse-drawn transportation of the Middle West. Oxen were brought in from the farms to haul wagons, and teamsters familiar with these bovine prime movers commanded fancy wages. Deprived of their horses, everyone in the sections affected by the epidemic was compelled to walk. C. A. Grasselli walked three miles morning and night from his home to the plant, while in Cincinnati Mr. Bailey must trudge seven miles twice a day. It was at this time that the high-wheeled velocipede first became widely popular.

During the first seventeen years of its operation there was no railway siding in the Cleveland plant. It was

Caesar Augustin Grasselli

Daniel Bailey

not until 1884 that a spur was run out from what was then known as the Valley Railroad. Up to that time all materials that could be shipped by water came into and went out of the plant from the adjacent canal. Horses and drays moved all other materials. The coming of the railway prompted shipments of acid in tank cars. These were built of iron plates and had a carrying capacity of 27,000 pounds. Remembering the 60 pound rails of that time these seemed heavy loads, but the Grassellis believed that greater capacities were possible, and in later years broke all records with a tank car carrying seventy tons of sulfuric acid.

Chemical demand having died during the panic of 1867, chemical prices sickened dangerously, and the convalescence of the market was made tedious and difficult because the largest consumers, the oil refiners, were just at that time engaged in a life-and-death struggle. In their fierce competitive battle each sought every advantage and all pounded away at sulfuric prices, endeavoring to purchase as cheaply as possible. The situation became critical for the acid makers. In self defense they formed a protective alliance.

In those days the approved method of thwarting the price chiseller was by means of "gentlemen's agreements," and Messrs. Eells and Schofield representing the Cleveland Chemical Works, David M. Marsh of Marsh & Harwood, and Eugene Grasselli agreed to hold down production and to stop cut-throat competition. They further agreed that Mr. Marsh was gradually to take over the Cleveland Chemical Works, and Mr.

Grasselli furnished a large part of the capital to make this purchase. Eventually these two acquired that company and so the ancient rivals became partners. This connection was more closely cemented when in 1870 they joined in buying out the plant built at Titusville, Penn., from Mr. Rainey of the Lodi Acid Works in New Jersey. Close to the Titusville plant they put a new refinery for the recovery of sludge acid. This plant was erected by John Metz, apprenticed a plumber, who became a chemical engineer and later was in charge of the works at Grasselli, New Jersey. Soon after this Eugene Grasselli bought out his partner's interest in these two Titusville plants and placed Julius Daub, a discreet Hollander, in charge. At the same time Mr. Marsh assumed active control of the Cleveland Chemical Works, having as secretary of that company I. H. Mansfield, formerly of Hussey and McBride, whose son Howard is today director of sales of Grasselli chemicals.

In the meanwhile the oil refiners were battling unmercifully. Prices went lower and lower and in 1872 all the important refiners gathered at the Metropolitan Hotel in New York. They invited the acid manufacturers from all over the country to meet with them. Their invitation veiled the threat to enter the acid business themselves so that the chemical manufacturers decided that discretion was the better part of valor and that they had best enter into negotiations. From their headquarters at the St. Nicholas Hotel they sent forth their envoy, Mr. Mansfield, to meet Charles Pratt, the minister plenipotentiary of the oil people. Mr. Mansfield delivered the counter-ultimatum that if the refiners made acid, the chemical makers would refine oil. At that time the chemical manufacturers, while not so numerous, commanded greater financial resources; and after protracted negotiations by this form of collective bargaining, which today seems very strange and unorthodox, a fair price for sulfuric acid was established.

From the very first the connections of the Grasselli firm with the development of the petroleum industry were intimate. In the very early days of the oil refining industry, Eugene Grasselli undertook in a small plant, known as the Newport Oil Company, Newport, Kentucky, to extract petroleum from cannel coal. For reasons not in the records this venture was never successful.

Many years later John D. Rockefeller, who had been an acid customer since 1860 and whose original refinery was almost adjoining the Grasselli Cleveland plant, proposed a combination of the interests to C. A. Grasselli. This proposal was undoubtedly prompted by the feeling that the chemical knowledge and experience then developed in the Grasselli organization could be made very useful in expanding scientifically the refining of petroleum. C. A. Grasselli was to take care of and develop the chemical industry in connection with the oil interests, which department was to be continued under the personal direction of Mr. Rockefeller. Mr. Grasselli inquired what provision was to be made for the Marsh interests, and receiving the reply that there was no place for Mr. Marsh in the picture, he refused because of his loyalty to his old friend to give the matter any further consideration.

The Cleveland newspapers of March 21, 1873, carried in their classified columns a modest announcement of great import to the Grasselli family: "My son, Caesar A. Grasselli, has been admitted as partner in the above works, said partnership to have effect from January 1st, 1873. E. Grasselli." The business was continued under the name E. Grasselli & Son, until the death of Eugene Grasselli, January 31, 1882.

C. A. Grasselli, or "C. A." as he was universally known among his friends and associates, continued the development of the manufacture of sulfuric acid as a major product. Early convinced that this acid ought to be produced close to its market, he considerably increased the number of Grasselli plants. The first steps in this direction were taken in 1889, when the Marsh & Harwood Company was wholly absorbed and The Standard Acid Works in New Jersey purchased. This latter was the nucleus around which the great plant at Grasselli, New Jersey, has grown. The Marsh & Harwood interests were already operating plants at Broughton, Pittsburgh, and Beaver Falls, Pennsylvania; Olean, New York; and Willow, Ohio. Expansion continued steadily. In 1892 the plant at East Chicago was built, and the plant at Birmingham, Alabama, in 1899. In this southern location the company branched out into the manufacture of acid phosphate and mixed fertilizers. Through the purchase of the

Standard Acid Company, Tonawanda, New York, in 1900 acetic acid was added and this material was later produced both at Grasselli and East Chicago. In 1902, with the purchase of the Standard Silicate Company, Fortville, Indiana, silicate of soda became one of the Grasselli products.

Up to this point the development of the Grasselli Company had been rapid but along established lines. Points had been carefully selected in the center of good consuming areas, and sulfuric acid plants were erected in which a logical chain of chemical products was produced. In 1904, however, C. A. Grasselli broke with this traditional policy. At Clarksburg, West Virginia, he erected furnaces for the manufacture of zinc. Additional plants for zinc smelting were later erected at Meadowbrook, West Virginia, and Terre Haute, Indiana. The extraction of sulfur from zinc-bearing ores was carried on in various sulfuric acid plants at East Chicago, Cleveland, Niles, New Castle and Canton, and the roasted cinders shipped to the zinc plants in Meadowbrook, West Virginia, and Terre Haute, Indiana. Just before his death another plant was added at Wurtland, Kentucky.

C. A. Grasselli was powerfully equipped for success as a chemical industrialist. He had, as we have seen, a thorough grounding in plant construction and operation. Behind this knowledge and experience he had a contagious enthusiasm for the chemical industry, backed by a profound conviction of its fundamental importance in modern civilization. He was an extraordinary executive, building up a huge organization on the model supplied him by his old friend, John D. Rockefeller. This was based on the committee form of administration, an organization to which the Grasselli Company adhered long after most large corporations had adopted the so-called staff-and-line system. His primary interest was always production, but he was a good merchant and he had great financial ability. During his lifetime the assets of the company under his control grew from $600,000 to $30,000,000.

His business life spanned the period of the American chemical industry's development from a comparatively few simple, standard, inorganic chemicals to the large scale production of a complex line of both organic and inorganic materials. His experience began at the acid chambers and ended at the head of the directors' table. Under his management the company passed from a personally conducted proprietorship to a highly organized corporation. Living through this tremendously expansive period, he himself grew. Yet to the very end he was always a personal leader rather than an impersonal executive. As long as he was active in affairs he maintained intimate contact with his men. And he was so cheery, so frank, so generous a personality that all who came in close contact with C. A. Grasselli loved him.

Like Edward Mallinckrodt, that other great chemical pioneer of the Middle West, he was an open-handed philanthropist. A devout Roman Catholic, his warm human sympathies reached out into those charitable institutions which administered to the sick, the helpless, the blind, the maimed, the orphaned. His closest interests were two homes, the one for the blind, the other for crippled children. In her later years his wife was an invalid, and after her death he remodeled and equipped their beautiful residence on Euclid Avenue and gave it as the Johanna Grasselli Home for Crippled Children. Another residence was given for work among the blind and is today the headquarters for the "Society for the Blind." Only a few months before his own death, "the blind" held a reception for him and presented him with a small silver cup which he kept on his library mantelpiece. "I keep it here," he said to intimate friends once. "It is too full of love even to hold flowers."

C. A. Grasselli married in 1871, a schoolmate of his sisters in Cincinnati, Johanna Ireland. On their wedding trip they went to Europe and then began long years of friendship and business association with a number of foreign chemical firms. First he went to Torno and saw on ancient doorways heavy brass plates bearing his family name. He visited the scene of his grandfather's earliest chemical triumph in Strasburg, which city was just then going through the phase of being assimilated by Germany after the Franco-Prussian War. He went to the Wohlgelegen Works near Mannheim and was shown the big, square, substantial stone building his grandfather had built, and in which he had made sulfuric acid from Sicilian brimstone. In addition to meeting the leading chemical manufacturers of France and Germany, he crossed the Channel and visited Sir Charles Tennant at his famous St. Rollox Works near Glasgow.

More than a quarter of a century passed before Mr. Grasselli again visited Europe. In 1899, accompanied by I. P. Lihme, the company's chief chemical engineer, he made another chemical tour. He visited the plants of Weiler Ter Meer near Cologne, of Vorstner and Bruneberg at Kalk, the zinc works of Julius Grillo at Oberhausen, the Frankfort and Griesheim plants of Cassella, the Bayer works at Leverkusen, the Goldschmidt zinc chloride plant, the Hoechst works of Meister Lucius and Bruning, the Merck pharmaceutical plant at Darmstadt, and the Badische Anilin und Soda Fabrik. Everywhere he made important business connections and won new, warm friends; Dr. Duisberg, Dr. E. Merck, Max Hasenclever, Dr. Pauli, Dr. Brunck, and Franz ter Meer. It is quite characteristic of him that after the World War he never wanted again to visit Europe.

C. A. Grasselli died July 28, 1927. Outside of the immediate realm of his chemical business he had won important distinctions. King Victor Emanuel III had knighted him in 1910 with the Order of the Golden Crown of Italy, and in 1921 made him a commander of that same order for the honor which he had brought to the name of Italy in other lands. In 1923 Pope Pius XI bestowed upon him the decoration of St. Gregory the Great. Two American universities had conferred upon him the honorary degree of Doctor of Science. For many years he had been president of two savings banks, which in 1921 were merged with the (Cleveland) Union Trust Company, of which institution he became and continued to be a director. Mrs. Grasselli had died in 1910, but their five children were living, T. S. Grasselli, president of the company, Eugene R., vice president and treasurer, Josephine and Ida Grasselli, and Mrs. W. T. Cashman.

A year after his death, in October 1928, the Grasselli and the du Pont interests were merged, and one hundred fifty thousand shares of du Pont stock, with a market value at the time of over $64,000,000, were exchanged. The consolidation was consummated by T. S. Grasselli and Lammot du Pont, whose fathers, C. A. Grasselli and Lammot du Pont, had, back in the early 80's, seriously considered a combination of their interests. Those negotiations had been abruptly broken off, March 29, 1884, by the sudden death of the elder Lammot du Pont.

C. A. Grasselli and the elder Lammot du Pont had many business dealings together, and this little personal footnote to the history of their companies is an appropriate ending to this story.

Lammot du Pont, who was a vigorous and original-minded chemical genius, visited C. A. Grasselli one day in Cleveland, seeking a sulfuric acid of then unheard of strength and purity.

"I think we can make it," said Mr. Grasselli.

"I'll bet you a box of cigars you can't."

The acid was made and delivered. In due course two boxes of the finest cigars were delivered to Mr. Grasselli's office. One of the original boxes lay in his porcelain-lined humidor, and in his handwriting on the bottom of the box was an inscription definitely identifying these as the cigars having been won from Lammot du Pont by the wager. On his death the cigar box and contents came to his son, T. S. Grasselli, who, when the consolidation was completed, gave them to Lammot du Pont, the son.

GEORGE T. LEWIS

1817-1900

EVERY industrial enterprise is born in the imagination of some man who conceives an opportunity to make some useful goods at a profit to himself. A century ago there were but few men who could visualize commercial opportunities in the manufacture of chemicals.

In truth such opportunities were then extremely limited. Furthermore, they were attended by more than ordinary risks of industrial pioneering. To venture into the business of making chemicals at that time, a man must needs be possessed of the reckless spirit of the gambler and the certain vision of a prophet. To succeed as a manufacturer of chemicals in those days of personal management required dogged perseverance and the soundest of good judgment.

If for no other reason than that they possessed this rare combination of talents, our pioneer chemical industrialists were all notable men. Not one of them was so distinguished by the ability to visualize a chemical opportunity as George T. Lewis.

He was our first great chemical enterpriser. He was associated with the first production in America of caustic soda. He was a leader in the commercial development of the Carolina phosphate rock deposits. He was one of the first refiners of cottonseed oil. He was instrumental in bringing the mineral kryolith from Greenland to the United States and employing it as a chemical raw material. Although he was the actual executive head of no chemical company, nevertheless he was the moving spirit behind three important chemical enterprises. Under his leadership the Lewis family interests in the production of white lead were expanded and diversified with other chemical products. To his experience and energy the New Jersey Zinc Company owes its existence. He was literally, as he was once called, "the Father of Penn Salt."

He was no chemist, yet with the devotion of a true scientist he spent his life in solving the economic problems of chemistry. Upon several occasions his absorbing interest in chemical development kept him from associating himself with other infant industries which have since grown to enormous giants. He turned his back on tempting opportunities to play an important and personally profitable rôle in the steel, coal, and petroleum industries, not because he was blind to those opportunities, but because he held fixed before him his vision of chemical opportunity.

This same dominating thought prompted him to hold the companies with which he was closely associated strictly in the chemical field. When difficulties arose, he sought a chemical solution of the problems. When expansions were planned, he resisted the temptation to branch out into new lines. His was always a thorough-going chemical policy, and these companies, as we know them today, are the fruits of the chemical seeds he planted and so carefully cultivated.

The focus of his attention upon chemical enterprises, sustained to the very end of his long life, must have been directed by some strong inner conviction. Certainly there was little in his family or educational background to have supported such whole-hearted devotion to the chemical industry. His schooling never included so much as a single lesson in chemistry, and in later years, despite his keen interest and heavy investments in chemical operations, he never saw fit to study a chemistry textbook or attend a chemical lecture. Naturally, so eager and industrious a man, closely associated with a variety of chemical operations, could not fail to pick up a considerable store of practical chemical knowledge, and almost unconsciously he became quite an expert in certain phases of chemical engineering, notably in smelter fumes recovery. Throughout his career, however, he depended upon others for his technical facts, and his concentration upon chemical enterprises certainly was not the result of an abiding preoccupation with chemical processes as such. Nor can the compelling fascination which the chemical industry exercised over him be explained as an inherited trait. It is true that his father and uncle had embarked upon the manufacture of white lead; but this chemical enterprise was always to them a side-line, a supplement, as it were, to their importing business. It was a venture in the spirit of the times when a number of the old trading houses were investing surplus funds in small manufacturing plants whose output they thus controlled and whose products they sold along with the multitude of miscellaneous imported goods they regularly handled. The Lewis family tradition was conspicuously mercantile, not chemical.

George T. Lewis was of the sixth generation in America of an ancient Welsh family, and from his father to his great great grandfather, four direct ancestors had been merchants and ship owners. William Lewis, first ancestor of the American branch of the family, came to this country in 1686, four years after the arrival of William Penn; and being himself a member of the Society of Friends, naturally settled in Penn's colony. He was descended from the illustrious family of Lewis of the Van, whose proud title traces back to Teon, the son of a long line of British princes, who at the close of the Fifth Century was Bishop of Coer Loew (now Gloucester) and later of London. From this city he was forced, by the invasions of the pagan Saxons, to flee to a refuge in the mountains of Wales. The same proud title, Lewis of the Van, is to this day borne by the stately ruins of the family seat in Glamorganshire in South Wales.

William Lewis, when he came first to Penn's colony, settled in Haverford Township not far from the present Wynnewood Station of the Pennsylvania Railroad.

Shortly he removed to a large farm in Newton Township. His son Evan Lewis, who was born June 7, 1677, left the farm of rolling acres in Chester County and moved to the growing town of Philadelphia. Here, in a modest little shop, he launched out in the mercantile line, to be succeeded in due time by his own son, Jonathan Lewis, who was born July 26, 1726, and who on Aug. 30, 1747, married Rachel Brentnell. Jonathan and Rachel Lewis had but one son, born September 21, 1749, and christened Mordecai Lewis after his maternal grandfather.

Mordecai Lewis greatly advanced the family fortunes. From small shopkeeper he grew into a great merchant, importer of foreign wares, exporter of colonial produce of all sorts, a ship owner and financier. He amassed a great fortune and built a fine house on Front Street, with his office and counting house just behind fronting on Dock Street. He was one of the prominent and responsible men of Philadelphia selected by his fellow citizens to countersign the paper money of the colony and later of the State of Pennsylvania. He took an active part in several of the public institutions of the city and in 1780 was elected treasurer of the Pennsylvania Hospital, to which post he was succeeded in turn by two of his sons and by his grandson, the office remaining continuously in the family for one hundred and one years.

Mordecai Lewis died March 13, 1799, at the comparatively early age of fifty-one, and he was succeeded by his son, Mordecai, Jr., who in 1806 took into partnership with him his younger brother, Samuel Neave Lewis, under the firm name of M. & S. N. Lewis. Three years later Samuel Neave Lewis married Rebecca Chalkley Thompson. Their fifth child, born August 3, 1817, was George Thompson Lewis who, like his illustrious grandfather, was named after his mother's father.

The young firm of M. & S. N. Lewis continued as commission merchants and ship owners in the business so successfully established by Mordecai Lewis, Sr. They prospered, and as young men are naturally inclined to do, began to branch out into new lines. As far back as 1772, their grandfather, Jonathan, had first imported white lead, and this valuable paint pigment had long been one of the important items on the manifests of the Lewis ships. In 1813 Joseph Richardson had begun to manufacture white lead in a small factory on Pine Street, out in the suburbs between Fifteenth and Sixteenth Streets. By 1819 his output had grown to the sizable total of a hundred tons a year and he began to attract the attention of the Lewis brothers. Seeing in him a threatening rival to one of their old and most profitable lines, they bought Richardson out, and so successful did the enterprise prove that by the end of ten years they had multiplied the production sixfold and by 1840 were manufacturing a thousand tons of white lead annually. As early as 1827, they started the manufacture of acetic acid (to take the place of the vinegar they were using in their process) and in 1830 they began to crush flaxseed and to market linseed oil to their paint-making customers.

George T. Lewis.

LEWIS' BAKING POWDER.

BEST IS CHEAPEST!

LEWIS' CONDENSED BAKING POWDER

STRICTLY PURE!

We will give $1000.00 for any Alum or other adulteration found in this POWDER.

Indorsed by the Brooklyn Board of Health, and by the best chemists in the United States.

It is **STRONGER** than **any Yeast Powder** in the world.

It **NEVER FAILS** to **make light bread** when used as directed.

It is **COMMENDED** by every housekeeper who has given it a fair trial.

It is an **entirely NEW INVENTION,** without any of the bad qualities of soda or saleratus, yeast or other baking powders.

It has in itself a tendency **to sustain and nourish** the system.

A single trial will **prove the superiority** of this Powder.

MANUFACTURED ONLY BY

GEO. T. LEWIS & MENZIES CO.

PHILADELPHIA.

Advertisement in the Illinois State Journal, Springfield, Ill., Dec. 5, 1879.

GEORGE T. LEWIS

It was shortly after these preliminary expansions of the white lead manufactory—in 1834 to be exact—that young George T. Lewis came to work in the counting house of the great firm of which his father was junior partner. At the time he was only sixteen years old; but he had completed the regular courses in the public and private schools of the city, and recognizing that he was not a natural student, his father acceded to his earnest plea to be allowed to enter business. His older brothers, John T. and Saunders Lewis, were already at work with the firm and like them young George was started at the bottom rung as a clerk in the accounting department. His strong natural bent for chemical enterprises soon brought him into the white lead factory.

The city of Philadelphia was growing fast. Fashionable Front Street where grandfather Mordecai had built his mansion had been usurped by business, and in 1816 M. & S. N. Lewis had torn down the old home and rebuilt an office and warehouse at 135 (now 231) South Front Street. The fine residential section was moving uptown along Chestnut and Walnut, Spruce and Pine, so that the full square block from Pine to Lombard, between Fifteenth and Sixteenth, where the white lead plant stood, was becoming altogether too valuable real estate for manufacturing purposes. The property was sold in 1848 and in 1849 the operation re-installed at Duke and Huntington Streets, in Port Richmond. An old white lead factory had stood on this site, but the plant was entirely revamped and enlarged. This work was all carried out under the direct supervision of George T. Lewis, and it was his inspiration that prompted the addition of such closely allied products as red lead, orange mineral, litharge, and acetate of lead.

The following year the old firm of M. & S. N. Lewis was formally dissolved. Both of the original partners had died, and the sons of Samuel Neave Lewis, who had inherited the business, agreed that the manufacturing and trading departments had so grown that they should be operated separately.

Accordingly, the firm of John T. Lewis & Brothers was organized in 1856 to take over the lead pigment business. Although he was the master spirit in this enterprise, George T. Lewis, with characteristic self-effacement, placed his brothers in active control, while he himself continued as commission merchant under his own name. It was agreed at the time that, in his capacity as commission merchant, George T. Lewis would sell the products of the Lewis lead factory.

This connection was the basis of other similar arrangements and was the true bond between him and a number of chemical enterprises. Because he sold packaged lye and baking powder, he became interested in the manufacture of caustic soda and alum, and it was his trade in fertilizers that suggested to him the organization of a company to mine the recently discovered phosphate rock deposits of South Carolina.

This venture, the Charlestown Mining and Manufacturing Company, was one of George T. Lewis' first and most successful chemical enterprises. He organized this company which soon became one of the largest in this field. Their own holdings in South Carolina were ten thousand acres of the very richest phosphate deposits, and they made mining leases for over twelve thousand additional acres. At the peak of the exploitation of the Carolina fields their annual sales were greater than a million dollars, and with the exhaustion of those fields they providently transferred their operations to Florida. Eventually the company was transferred to interests allied with the Virginia-Carolina Chemical Co. fertilizer merger.

As early as 1847, George T. Lewis brought cottonseed from the South to his own mill in Philadelphia and began crushing it. It was his belief that cottonseed oil would become a substitute for linseed oil, and when the paint trade could not make this substitution instead of abandoning the project, he engaged a chemist, worked out a method of refining his cottonseed product, and sold it for use in oil-burning lamps and in soap making.

While in close personal touch with the manufacture of white lead, he proved himself one of our first chemical conservationists by working out a successful method for the recovery of lead in the smelter fumes that had formerly gone to waste. As was his custom, in attacking this problem he associated with himself a good technical man; but the practical experience he gained in this way at first-hand enabled him years later to regenerate the Lehigh Zinc Company. He furnished plans for the first successful zinc furnace built in the United States and later built for the New Jersey Zinc Company a furnace for smelting the rich, but difficult Franklinite ores.

Just at the time of the dissolution of the old firm of M. & S. N. Lewis, Dr. Richard Tilghman, a prominent physician and member of the distinguished Eastern Shore family, came to Philadelphia from Baltimore. He had a new chemical idea expressed in tangible form: a patent covering a new process "for the manufacture of the alkaline salts of soda." George T. Lewis, who at the time was on the outlook for new business opportunities, and Charles Lennig, the aggressive proprietor of the largest heavy chemical works in Philadelphia, were naturally among the first wealthy men the inventor approached. They were both favorably impressed by the doctor's representations and the commercial possibilities of his process. After a long series of conferences, they agreed to buy the Tilghman patent outright for twenty-five thousand dollars.

Meanwhile other important Philadelphians had become interested in this proposed chemical enterprise, and it was finally agreed to organize a new corporation to acquire the patent and work the process. Accordingly, on September 25, 1850, a charter was obtained "for a period of ten years and for the purpose of manufacturing salt and its resultant products" by a company with twenty shares of five thousand dollars each, all subscribed and paid in. Thus, the Pennsylvania Salt Manufacturing Company was launched.

The original board of directors, consisting of Charles Lennig, president, George T. Lewis, George C. Carson, Samuel Sims, and Samuel F. Fisher, with George Thompson as secretary and treasurer, drew up a carefully planned program. Of their one hundred thousand dollars working capital, a quarter was by agreement invested in the Tilghman patent. They next assured themselves of control over their raw materials by purchasing twenty-nine acres of land with proved salt reserves underground and including a coal privilege 500 feet on the front line and one mile in depth. At the same time they took an option on additional coal privileges of 150 acres at thirty dollars per acre. For this property they paid $5750.00, leaving them as working capital $69,250.00. Subsequently their cautious investment in land proved costly, for the Company was later forced to pay up to as high as six thousand dollars an acre for land that they might then have purchased for not more than fifty dollars. However, had they bought a couple of thousand acres, as their successors devoutly wished they might have done, either the original investment must have been doubled or else the available cash would have been so depleted that it is doubtful whether the enterprise could have struggled through its first seven very lean years.

The situation as they saw it, and as it developed during the early life of the Company, fully justified their conservative investment in land; and they were certainly more than justified in the caution with which they proceeded in their manufacturing operations.

With a new process to translate from patent to plant, they decided to entrust the experimental work to experienced Charles Lennig, and before building their own plant to employ his existing manufacturing facilities. Having tied him tightly with an extremely carefully drawn operating agreement, they turned the process over to him, and he took it out to his works at Bridesburgh for trial. At the same time they sunk wells, installed brine pumps, and erected the outer frame of their plant buildings on the newly acquired land at Natrona, near Pittsburgh.

Charles Lennig could not produce the alkaline salts of soda according to the specifications of the Tilghman patent. For two years he tried to do so, and then after some rather costly modifications of apparatus, the process was transferred to the new Natrona plant. The process could not be made to work. It proved a discouraging and costly failure. In 1854 the directors resigned. Dissolution of the corporation was freely discussed. But the old board was re-elected, and it was at this low point in the corporation's affairs that George T. Lewis earned his title of "Father of the Salt Company."

He was insistent that they find a chemical solution for the problem. As a stop-gap the company had been selling salt; but he pointed out the ruinous competition from the Ohio Valley wells and convinced his associates that their future lay not in salt, but in "its resultant products." Investigations were started into the use of salt as a chemical raw material, especially looking to the production of soda ash, which were later described as "the foundation of our business and security for further proposed investment."

In 1856 the first net profits, a modest $1047.58, were earned, and that same year a patent was secured for George Thompson's famous "Saponifier." So convinced was Mr. Lewis of the ultimate success of the enterprise and of the value of this Thompson patent, that he not only advanced the better part of $100,000 for additional plant and continued chemical investigations, but he also agreed to take over the sales of the "Saponifier."

The substitution of the cheaper caustic soda for caustic potash, which had previously replaced the messy and unreliable hardwood ashes in the home manufacture of soap, was a revolutionary household convenience that today we cannot appreciate. Seventy-five years ago, even to the majority of city dwellers, a cake of hard soap bought in the store was a luxurious toilet article. Most of the cleaning and laundry soaps were homemade; and more than that, packaged lye, before our multiplicity of water softeners and scouring powders, was the popular cleansing agent. It became the basis of a tremendous business, the foundation of such familiar chemical trade-names as Ford and Babbitt, and it supplied the beginnings of the Guggenheim family fortunes. From Ford has sprung the Michigan Alkali Company. The Babbitts went on to the manufacture of soaps. The Guggenheims sold out their famous "American" brand of lye to George T. Lewis and on the advice of one of the sons invested it in Montana copper-mining operations. But from the earliest days till now Lewis Lye has been a leader in this field.

Allied with the sale of packaged lye in his business as commission merchant, George T. Lewis carried alum baking powder and disinfecting chlorides, products which were made for him by the Pennsylvania Salt Manufacturing Company. It was his natural interest in alum that turned his attention to that curious mineral kryolith, a double sodium and aluminum fluoride found in commercial quantities only in one vast deposit on the south shore of Greenland owned by the Danish Government. This single commercial deposit is of enormous extent, the main mass being six hundred feet in length, two hundred in width, and of still unknown depth. Moreover it is of exceptional purity, whole cargoes sometimes running over 98 per cent. pure kryolith. About 1850, a young Danish chemist, Julius Thomsen, worked out a commercial process for the use of this mineral as a chemical raw material by decomposition with lime, the primary products being caustic soda, soda alum, and calcium fluoride. In 1864, King Christian IX granted a mining monopoly to the Danish Kryolith Corporation, and they in turn supplied the material to the chemical firm of Weber & Brothers of Copenhagen under whose auspices the researches of Thomsen had been conducted. The very year that these Danish agreements were signed, Mr. Lewis brought the matter of kryolith to the attention of his fellow directors and proposed that Henry Pem-

Photograph of George T. Lewis made in June, 1869.

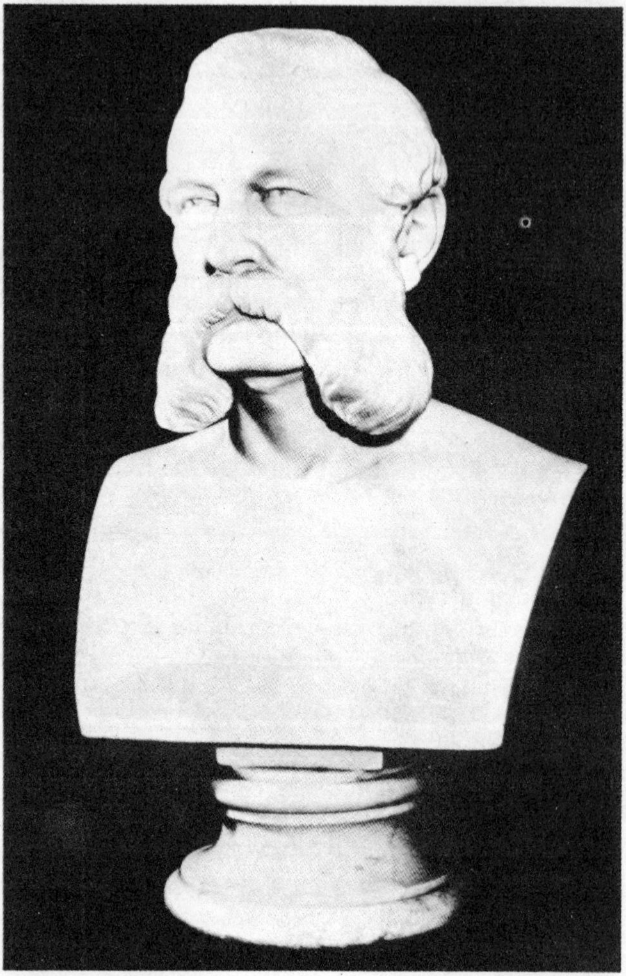

Bust of Mr. Lewis completed about 1900.

berton, their plant manager, and his son, Samuel N. Lewis, be sent to Denmark commissioned to negotiate an exclusive American importing contract for this interesting material.

The following spring the two agents went to Copenhagen. After four months' investigation and negotiation, they entered into an agreement with the Kryolith Corporation which has remained in force continuously to this day. The original contract called for a minimum annual importation of 6000 tons, and during the life of this agreement importations have several years exceeded four times this amount. As a result of securing this new raw material, the Pennsylvania Salt Manufacturing Company, in 1867, put on the market the first American-made bicarbonate of soda and from the same source has recently been able to commercialize modern discoveries in the use of fluoride insecticides.

In 1873 George T. Lewis resigned as a director and so terminated his active connection with the management of the Pennsylvania Salt Manufacturing Company. At the same time he turned over to the corporation, upon a favorable royalty basis, his famous brands of lye. These were widely and favorably known, for not only had he been a pioneer in the introduction of caustic soda in this field, but as a shrewd merchandiser he had improved on the solid cake by furnishing first the chip and later the powdered form. He had also developed not only the first tin can but improved the closure, first with a pouring spout and later with a sifter top.

It was preeminently a characteristic of George T. Lewis' business methods that he furnished the ideas and then called upon others to work out the details and to carry them forward to successful execution. An extremely successful merchant, he was no salesman; a great chemical enterpriser, he was neither chemist nor executive. His fertile brain was the dynamo that supplied the power which his associates and employees applied to useful, profitable purposes. His natural habit was to delegate authority and to place responsibility squarely upon the shoulders of his associates. His methods were successful, because he was a good judge of character and capability and also because of his own urbane temperament and sense of justice. He was genial, friendly, witty; and he had the gift of dealing with either his most dangerous competitor or his own office boy upon a personal basis without the loss of his own dignity or the jeopardy of their wholesome respect. He was too honest and honorable to be sharp. But he was keen and shrewd. He believed thoroughly that the first object of business is to make a good profit, and even today it is remembered in Philadelphia that he was without a peer in analyzing a cost sheet or a quarterly statement.

Like other distinguished and wealthy Philadelphians of his day, he bore an active part in the semi-public and charitable institutions of his city. His particular favorite was always the Church Home for Children, which he served faithfully as treasurer for more than a quarter of a century. During the Civil War, he helped to organize and equip the "Corn Exchange Regiment" (118th Pennsylvania Volunteers); and though he was never an office seeker, he long carried an important, unobtrusive part in civic affairs, where he was a power for good and honest government.

George T. Lewis died January 17, 1900, aged eighty-three, the sole survivor of incorporators of the Salt Company; his wife, Sally Fox Fisher, whom he had married May 18, 1843, and four of their five children surviving him. Sixty years previous, the obituary of his own father, published in the "National Gazette" of February 11, 1841, might well serve for himself: "Few men have passed their lives more usefully and less obtrusively . . . . Educated as a merchant, with the favorable principles which distinguished his ancestors, he soon became one of the brightest ornaments of our commercial circle. His unassuming nature did not permit him to be much before the public, although his talents, especially as an able accountant and skilled financier, ever impressed all with whom he was engaged in business."

LUCIEN C. WARNER

1841-1925

SEVEN months' hard labor had come to a second withering failure. William D. Patten, two chemists, and an engineer had slaved day and night to devise a process that was quicker and took less plant space than the old leaching tank operation for producing phosphoric acid. A tidy sum had been laid out in apparatus. Precious time had been lost. And as yet a solution of this problem—vital to the plans for expanding the young Warner Chemical Company—was as far away as ever.

The sorry record was laid before the president. He listened quietly, very attentively; and carefully he checked over the costs of the experiment.

"Now, Mr. Patten," said Dr. Warner, shoving the stack of papers deliberately aside, "what have we learned from this failure?"

No word of reproach, no regrets for time or money wasted, no swerving from the approved plan; only quiet determination to win from disheartening failure a new and better way of again attacking the same old problem.

That anecdote aptly illustrates the characteristics that enabled Lucien C. Warner, at the age of fifty, to become a chemical manufacturer and to succeed in this complex, technical business.

It is a story almost in the vein of Oliver Optic, for the very next trial solved the phosphoric acid problem by means of a system of vacuum filters. It is quite in the approved chemical style to add that this hard-won process was itself shortly to be replaced by the Dorr system of counter-current filtration. But in the meanwhile the development of The Warner Chemical Company had gone forward apace.

A quarter of a century before Dr. Warner had formed that wholesome habit of profiting from his own mistakes. At that time, a young man of twenty-six, just graduated from the Medical School of New York University, he and his brother were traveling about the smaller cities of northern New York delivering a course of popular lectures on health and hygiene. In developing a technique of successful publicity, he learned his first practical business lessons. Upon reaching a new city, it was their custom first to give a free lecture; and they soon discovered that what attendance they could muster to this sample performance very accurately gauged the number of tickets they would sell for the subsequent series of five talks. If their free demonstration came into direct competition with some popular local event, or if the evening they had chosen should prove to be stormy, their entire stay in a town was apt to be unprofitable. Years later when Dr. Warner published his reminiscences in "Seventy Eventful Years: 1841-1911," he wrote: "I feel that these bad towns were the most profitable of all, as they taught us how to meet obstacles and wrest victory from defeat."

This spirited tenacity, which he retained till the very day of his death, certainly stood him in good stead, when late in life, without experience in the chemical industry, and with only the sketchy knowledge of chemistry that was a prerequisite for an M.D. degree received in 1867, he took over a concession from the French Government to exploit a rock island off the Guiana coast of South America. He had on his hands a rich, but intractable, chemical raw material. It took a dozen years and a deal of clever technical juggling with processes and products to convert it into a profitable chemical enterprise. If he had no formal training for chemical manufacturing, Dr. Warner had certainly been well schooled by the hard taskmaster, Experience.

From both of his parents, through all lines of ancestry, Lucien Calvin Warner was descended from English stock who settled in New England prior to 1660. His father, Alonzo Franklin Warner, was of the eighth generation from Andrew Warner who came from England to Cambridge, Mass., in 1632. This first of the American Warners must have been an incorrigible pioneer, for five years later he moved westward to the frontier town of Hartford, Conn., and again, in 1659, on further to Hadley, Mass., where he died in 1684. For five generations the Warners stayed in western Massachusetts, and the town records of Hatfield, Plainfield, Enfield, Cummington, and surrounding places are marked well by their distinguished services against the Indians, in the Revolution, in church, school, and town offices. Then Ira Warner was again stirred by the westward urge to leave Plainfield, and with his wife, Asenath Hitchcock, he settled between Cuyler and Lincklaen, in central New York. The eldest of the twelve children of Ira and Asenath Hitchcock Warner was Alonzo Franklin Warner, father of Lucien.

As eldest son, Alonzo, save for a couple of years when he taught in the local school, was his father's chief helper on the farm. In return he was given as a wedding present a farm of about a hundred acres, near Lincklaen, a scant mile from the family homestead, and here, on December 24th, 1838, he brought his bride. She was Lydia Ann, the daughter of neighbors, Calvin and Harriet Fuller Converse. Through her father she was the descendant of Deacon Edward Converse, who came to Massachusetts with John Winthrop in 1630, and having settled in Charlestown established the first ferry to Boston, which he later turned over to his friend John Harvard in order that the income from this early public utility might be devoted to the support of Harvard College. Through her mother she traced her ancestry straight back to Captain Edward Fuller, one of the *Mayflower* passengers who became leader at Plymouth and Barnstable. Of this rock-ribbed lineage Dr. Warner himself wrote: "The family were all industrious and respected citizens, and nearly all were abstainers from both tobacco and liquor. The men were all farmers, and the women all married either farmers or tradesmen, excepting one, who married a physician. All were in comfortable circumstances, but only one, Lorenzo, manifested any special ability in making money. They belonged to that great class of self-respecting, intelligent and thrifty citizens so common in the early history of our country among the descendants of the early New England families."

Lucien Calvin Warner was born at the old family homestead near Cuyler, N. Y., October 26, 1841. His brother, Ira De Ver Warner, had been born in the new farmhouse at Lincklaen, March 26, the year before. Shortly before the birth of the oldest boy, the grandfather had been kicked by a young horse he was breaking and had died. Accordingly, the father, as oldest son, had been forced to return to his old home to take care of his numerous younger brothers and sisters. The management of this estate and the care of the big family were heavy drains upon his strength and resources, and within a few years, when his own sons were but seven and five, he died suddenly of pneumonia. His widow was just able to salvage enough to purchase a poor, little farm near Lincklaen, and here by dint of hard work and careful management she brought up her boys. She made most of the clothes, and they grew most of their own food. The only cash income, and this seldom exceeded a hundred dollars a year, came from the sale of butter from two or three cows. It was "a

Lucien C. Warner

Lucien C. Warner and his older brother, Ira Der Ver Warner, from a daguerreotype taken when they were schoolboys.

life of extreme frugality, though not often of actual hardship."

As a young boy Lucien planned to go to college. At the local school he learned quickly and remembered well, so that by the time he was sixteen he was himself teaching in a district school during the winter and attending the spring and fall terms at the De Ruyter Academy. The headmaster, Rev. Shubael Carver, encouraged him, and through his influence a scholarship was obtained at Oberlin College which young Warner entered in August, 1860. Here the same course of teaching during the winter term in order to earn his way through the spring, summer, and autumn semesters was followed, but even this had to be supplemented by serving as carpenter and handyman for the college at the wage of ten cents an hour. He paid a dollar a week board at a students club and fifteen cents a week for his unfurnished room in the dormitory. His total expenses the first full year at Oberlin were $160, about $75 of which he earned by manual labor, the balance by teaching.

These were the stirring days of the Civil War, and twice young Warner forsook college for brief tours of military service. In the summer of '63 he volunteered with a company of fellow-students to defend Cincinnati, threatened by General Morgan's brigade; but by the time they reached the city the Confederates were already in retreat toward Kentucky. In the spring of '64 he answered a call for volunteers to man the forts at Washington in order that the veteran troops held there might be released to join Grant's drive upon Richmond. Here he smelled powder burned in action, being a corporal in charge of one of the eight-inch cannon at Fort Stevens which was the focal point of General Early's abortive raid upon the capital.

August 24th, 1865, he was graduated from Oberlin. The next two years found him studying medicine at New York University. For the next six years the young physician practiced medicine in McGrawville, N. Y.; traveled in partnership with his brother delivering those lectures of health and hygiene which gave him his first real business training; and published two popular medical books, "Functions and Diseases of Women" and "Man in Health and Disease." In 1873 he came to New York to practice medicine; but the following year, entered into a new partnership with his brother for the manufacture of corsets.

The success of the Warner Brothers corset was immediate and very great. Within ten years the sales passed a million dollars annually and by 1892 reached a total of $2,691,264. His brother had complete charge of the manufacturing in the factory at Bridgeport, and Lucien Warner in New York looked after the buying, the advertising, and the selling. Although this was his chief business interest and provided the surplus funds which led him into the chemical industry, nevertheless this is not the place to tell again the story of the W/B corsets.

Like many wealthy men Dr. Warner sought to use some of his capital in a number of ventures outside his principal business. Most of these turned out rather badly, and late in life he set forth an epitome of what he had learned by his mistakes:

"Most failures in life grow out of the unwillingness of a person to profit by the experience of others. My own mistakes would nearly all have been avoided by observing four short rules:

Rule 1—Never loan a needy friend any more money than you can afford to lose.

Rule 2—Never invest in a business with which you are not familiar without first having a careful investigation made by disinterested experts of established skill and responsibility.

Rule 3—Never take stock or invest money in any new enterprise unless you are willing to devote to it your own time and energy.

Rule 4—If you are conducting a successful business do not add to it any other business, unless so closely related to it that the one helps the other."

The Warner Chemical Company proved to be the shining exception that, if it did not prove all rules, did at least disprove Rules 2 and 4.

One of Dr. Warner's unsuccessful ventures was a wall-paper company through which he became acquainted with Harris H. Hayden, an ingenious and over-optimistic soul, who with Walter S. Pearce had become interested in deposits of phosphate of alumina on Grand Connetable Island, near French Guiana.

J. Swan & Company had a concession for mining this phosphatic rock which they had transferred to the Grand Connetable Company, which in turn had assigned mining rights to the International Phosphate Company.

Among the stockholders in this last corporation was Sanford H. Steele, the attorney who later was to be vice-president of the General Chemical Company, and following Rule 2, Dr. Warner sent Mr. Steele down to the rocky islet of the South American coast to examine and report on the extent of the deposit. He estimated that, as near as he could judge, there were about three hundred thousand tons of rock, an estimate that eventually proved to be just one-third too large. The rock analyzed high in phosphoric acid, but being combined with alumina, was recognized to be difficult to make available. Nevertheless Dr. Warner was convinced that there was great value in this rock and he took an interest in the company. "One reason was that it seemed an opportunity for a profitable investment; but the stronger reason was the natural desire to do something that would compensate for the loss we were sustaining in the wall-paper company."

These may not have been very sound and logical reasons for embarking upon a career as a chemical manufacturer. They were, however, frankly faced, and once in chemical making Dr. Warner put into force Rule 4 and devoted his own time and energy to the project. The first effort was to extend the market. Up to that time the chief sales had been made in bulk to fertilizer manufacturers. Dr. Warner quickly sensed an uncertain future in this field. The phosphoric acid was not in a soluble form and the value of the material as a plant food was already questioned by chemists. A search for new uses was undertaken in the wall-paper factory, but shortly a plant on the Harlem River in the Bronx was leased and experimental work transferred there. The first effort was to combine the rock chemically with ammonia and later with potash for fertilizer use. These trials came to nothing practical. Tri-sodium phosphate was next attempted, and Mr. Hayden, assisted by a young chemist, finally found a new, workable process.

Encouraged by this success, Dr. Warner in 1891, when he was fifty years old, purchased the stock of Mr. Steele and took over from the J. Swan & Company their interest in the Grand Connetable Company which carried with it the original concession from the French Government. These holdings gave him control, and in 1897 he bought the stock owned by Mr. Pearce, so that he and Mr. Hayden became the sole owners. The following year Mr. Hayden retired from the business in order to go with the Eastman Kodak Company, and Dr. Warner bought his one-third interest.

Dr. Warner decided materially to expand the business. His son, Franklin H. Warner, who had evidenced a lively interest in chemistry at college, would be entering business shortly, and would eventually take over the administration of this chemical enterprise. Instead of expanding the little plant on the Harlem River, another waterfront location was found (in order, of course, to facilitate the shipments of raw material from Grand Connetable Island) and a new plant erected at Carteret, N. J. This site, at that time an undeveloped semi-swamp, bordering Staten Island Sound, has since become a great center of chemical manufacturing activity. Its selection by Dr. Warner is good evidence of his business acumen and foresight in a problem to which he devoted his personal attention, for he had now become a full-fledged chemical industrialist. "Up to this time," he wrote in his brief autobiography, "I had given but very little personal time to the enterprise . . . . but I plainly saw that if I enlarged the works and went on with the business, I must give it more personal care. This I hesitated to do, and yet there was a fascination about the unsolved problems of chemistry that attracted me."

The larger plant soon meant a longer list of products and to tri-sodium phosphate and phosphoric acid were added the di-sodium salt, in both the U. S. P. and commercial grades, for which a fast-growing market was found in the then-new process of silk weighting.

The new plant had been operating a year and this new product was just getting well established when Dr. Warner received a visit from a young man with a new chemical idea. Both patent and patentee were to become extremely important in the subsequent history of The Warner Chemical Company. The man was William D. Patten. The patent was his basic one upon the use of acid-sodium pyrophosphate for a baking powder.

Young Patten, who had been manager of the soda business of Alex J. Howell, had discovered the superiorities of the acid-sodium pyrophosphate for this use, and he was naturally attracted to the rapidly expanding Warner Chemical Company which was obviously well placed as to raw materials and had so recently given evidence of chemical progressiveness in the field of the phosphate salts. The acid pyro salt was at that time selling at a dollar a pound. If this baking powder market were to be developed, it must be sold at least as cheap as cream of tartar. This meant reducing the price at least three-quarters. Both commercially and technically the crux of the matter was a cheaper method of manufacturing the pyrophosphate.

Dr. Warner asked a lot of pointed questions which revealed that he was fast learning the chemical business.

Mr. Patten answered fully and frankly and with no intention of attempting to interest him in the project, for at the time all that he was seeking was a reliable and reasonable source of his raw materials.

Finally Dr. Warner asked point blank, "Mr. Patten have you enough money to develop this idea of yours?"

"I have enough money, I think, to work out the cheaper method of manufacturing. When that is done, knowing that the product is a better baking powder, it will then be the right time for me to interest capital. I am not going to try to raise the money for developing the market until I have my project on a commercial basis. Then I should have little trouble securing capital for expansion."

"I see," Dr. Warner commented drily with a smile.

On Christmas Eve, 1901, Dr. Warner sent the message that if acid-sodium pyrophosphate could be made for ten cents, his company could do it, and that he would like to talk business.

From the very first these two men were attracted to each other and the commercial arrangements were very promptly made. The Warner Chemical Company was to make the acid-sodium pyrophosphate for two new selling companies, the Monarch Chemical Company in which Patten's interest was two-thirds and the Warner Baking Powder Company in which the two-thirds interest was held by Warner. The distinction was based upon recognition of two different merchandising problems to be solved: sales in bulk to bakers and sales to housewives in small packages. Monarch's bulk business prospered. The household market was never captured. In the end the Warner Baking Powder Company was dissolved, and it is typical of the relationship that always existed between these two that Mr. Patten voluntarily readjusted their stock interest in the successful Monarch concern to a 51-49 basis.

The first step after this affiliation was to build a plant for the manufacture of acid-sodium pyrophosphate. The operation ran into a number of anticipated difficulties and developed some that had not been foreseen; but at the end of a year, during which time Mr. Patten literally lived at the Carteret works, the kinks had all been smoothed out of the process and the low cost objective won. This success stimulated Mr. Patten to per-

Mrs. Keren Osborne Warner

Judge Noah H. Osborne, Mrs. Warner's father, with the Warner children. Left to right, Elizabeth (Mrs. William G. Gallhower), Lucien C. Jr., Franklin Humphrey, and Agnes (Mrs. Seabury C. Mastick).

fect a new process for the production of the mono calcium phosphate, also used in baking, and led shortly to a German connection that opened up a profitable foreign market and brought the Neuberg family into connection with the Warner enterprises.

A resourceful German chemist of the staff of the Goldenburg-Geromont Company of Wiesbaden, Dr. Oscar Neuberg, had developed a means of using tartaric acid as a leavening agent, and he came to America to exploit this process. Hearing of the phosphate baking powder development he went straight to the Warner offices, and after protracted negotiations, entered into a contract whereby his German company were to act as exclusive selling agents. Through Dr. Neuberg's introductions similar profitable connections were established in Great Britain. His brother, William, was an independent sales representative in New York for German chemicals, and two of his nephews, sons of William Neuberg, became in time the heads of two important sales divisions of the great corporation which has grown out of the Warner Chemical interests.

While working on these phosphate salts, Mr. Patten developed a process for utilizing the alumina of the Grand Connetable rock in the form of aluminum hydrate, thus turning a waste into one of the most important of the Warner chemicals. At the same time Dr. Warner was shoving along his program of diversification in other directions. Vanillin and salol were produced; but fitted illy into the chain of the chemical operations at Carteret. They were promptly abandoned, for it was one of Dr. Warner's favorite sayings that "it is the greatest indication of courage to stop when you are on the wrong course."

Out of the vanillin mistake, however, was salvaged the greatest triumph of all. Among the liquidated assets of this experiment were some chlorine electrolytic cells developed by a young chemist, H. R. Nelson, who begged Dr. Warner to allow him to go on with this work. In due time, an economic source of pure chlorine resulted in the announcement of such new and highly logical products as phosphorus trichloride, phosphorus oxychloride, acetyl chloride, acetic anhydride, and carbon tetrachloride. When the World War came, Warner was the sole American producer of these phosphorus-chlorine compounds and acetic anhydride which sprang to immediate, vital importance both in munitions making and in dye synthesis. Furthermore, it was through carbon tetrachloride that Dr. Warner and Ernest C. Klipstein were brought together.

They quickly found several common chemical interests interlinking with the electrolytic chlorine-caustic operation. Klipstein sorely needed a domestic supply of carbon tetrachloride for his Carbona company. Moreover, his new sulfur black plant required caustic soda and he was planning the manufacture of anthraquinone and would soon need chlorine. Warner not only used caustic in the initial steps of producing his various sodium phosphates; but he had a dozen good uses for more chlorine. Over and above all, there was a big demand, at war prices, for both these basic chemicals, so that any surplus above intra-plant consumption was sure of a ready, profitable sale.

They organized the Warner-Klipstein Company, gave Nelson a block of stock in recognition of his development of the electrolytic cells, selected South Charlestown, W. Va., because of its coal and salt, and built there a new plant. This operation was continuously and rapidly expanded throughout the war period. At the same time The Warner Chemical Company entered the equipment business and erected more than twenty similar electrolytic plants all about the world from Norway to India, from Australia to Canada. After we entered the war, they built for the Chemical Warfare Service at Edgewood the largest electrolytic chlorine plant that up to that time had been constructed.

When, after the Armistice, the enormous, but artificial demand for all sorts of chemicals stopped abruptly, Dr. Warner boldly cut and consolidated the operations in both plants. He took over the common stock interest of E. C. Klipstein and all the interest of H. R. Nelson, who elected to take a cash settlement and retire from active participation. Having trimmed the ship, he stepped down from the bridge, and his son Franklin H. Warner succeeded him as president.

Dr. Warner died in 1925, and three years later the Westvaco Chlorine Products Corporation was organized to merge into one unit the various Warner chemical enterprises. Twelve years before a young man, William B. Thom, had joined the organization, as assistant to Franklin Warner who was then the treasurer. When the young Mr. Warner retired to California, the younger Mr. Thom became the active head of the recently consolidated corporations. Though he thus became the most youthful of all big chemical company presidents, he disclaimed that youth must be served.

"Dr. Warner was the perfect refutation of that silly saying. He himself said that interest and energy, because they combined in hard work, were the prime requisites for success, nevertheless he was a living example of the value of experience. He was eighty-three years old and he had made two fortunes in two very different industries; but he was keen and alert, open-minded to new ideas, enthusiastic for new ventures. His practical, first-hand business experience was very great, and nothing can replace it, for it is the surest guide to correct policy and right action."

Lucien Warner possessed mental traits that in the law would have led him to the judicial bench. He had a rare gift for listening to the pros and cons and then summing up the question in a few terse, clear sentences. Even today his associates are fond of quoting many such "Poor Richard Proverbs" of his as:

"If you have several reasons for not doing something, pick the best and give that one only."

"If you are not in business to make money, you will soon be put out of business."

"A committee is slow, but it is sure."

As quite a young man Dr. Warner consciously chose to devote but half his waking hours to business, and no picture of his life would be truthful that omitted two great outside interests, foreign travel and public works. When he was working, he was in the office before the first mail and even in his latter days put in a full day at his desk. He simplified his routine and concentrated his efforts masterfully, never slighting any task or responsibility; but deliberately planning to restrict his business hours. For many years he was a devoted friend of the Y. M. C. A., Oberlin College, and the Congregational Church, giving generously of his wealth and his time to these causes that had his warm sympathy and active support. Since 1880, when he and Mrs. Warner made their first trip abroad, he went on long trips every year. This was his great hobby, shared with the wife who was constant companion and closest comrade.

At seventy, after careful thought, Dr. Warner reached a decision that he kept to the end: "As a physician," he said on his birthday, to one of his sons, "I am aware if I watched my diet and my exercise and was careful to dodge away from winter cold and summer heat, it is likely that I should be able to add several years to my life. If I did so, I should have time and energy for little else, and I should quickly become quite useless to other people. I prefer to go on with my works, to do all that I can as long as I am able."

The thirteen years he lived after reaching this decision were momentous ones for the chemical industry since they embraced the period of the World War.

They were most important years in the history of the business that we know today as the Westvaco Chlorine Products Corporation. The courage and the tenacity and the great business experience of that sturdy veteran, who had resolved to carry on during these strenuous times, made notable contributions to the growth of our American chemical industry.

EDWARD MALLINCKRODT

1845-1928

JUST prior to the outbreak of the World War business activity in the United States was at a low ebb. Even in the fine chemical industry, where the medicinal trade is always a stout backlog, orders fell off and stocks began to pile up in the warehouses of the Mallinckrodt Chemical Works.

This growing inventory disturbed Edward Mallinckrodt, the owner of the business, less than it did a young chemist who had but recently been transferred from the laboratory to the purchasing department. Keenly interested in his new work, he compiled figures showing how much production should be curtailed to balance with shipments. So armed he sought out the opportunity to lay his facts before his employer.

Edward Mallinckrodt listened attentively and checked over the carefully arrayed columns of dollars and pounds. He made an inquiry or two, even threw out a couple of suggestions, and finally asked, "Well, what do you propose I do?"

"With a few exceptions we have stocks sufficient for six months, and I suggest we virtually close the plant—just keep a skeleton force at work. You can see, Mr. Mallinckrodt—right there," and the young man pointed to his figures, "the saving would amount to several thousand dollars a day."

"So it would, so it would," nodded Mr. Mallinckrodt. He leaned back suddenly, and looking up at Frederick Russe, smiling, he said soberly, "Sit down, I want to talk to you about this. Russe," he asked, "times are pretty bad in St. Louis, aren't they?"

"They are, indeed."

"If we discharged several hundred of our people, do you think they could get work elsewhere?"

"No, I'm afraid—"

"That would cause a good deal of hardship, even suffering in some cases, wouldn't it?"

"Well, yes, I dare say some of—"

"If we keep them all at work I am the only one who suffers. I own this business. There are no stockholders to be taken care of, and if we lose a few thousand dollars it comes only out of my pocket. I can afford that better than they can. They don't have a lot of money: the least I can do is to see they do have steady work."

There was no answer to that argument, but he continued quickly as if to erase any possible suggestion of parentalistic philanthropy.

"Remember, too, Russe, that just as we don't have good times all the time, so we won't have bad times for-

Plant of the Mallinckrodt Chemical Works in 1868—original buildings were the oblong stone factory and adjoining wooden shed in the rear.

ever. Before long we shall be able to sell all the stocks we are now making up."

With the World War came an over-demand for all fine chemicals, so that within the year the stockrooms were stripped bare of that threatening surplus. Edward Mallinckrodt had no occult gift which enabled him to forecast that world catastrophe, but it was his natural habit to scan both the immediate details and the long range effects of every business proposal. Thus the point he stressed during bad times has its counterpoint in an incident of the booming war days.

To handle the extra business, the plant force was working double-shifts, and the clerical staff had been so increased that the desks of typists and order clerks were sandwiched into the big administrative building till there was scarcely aisle space left. To a number of the younger department heads it seemed an opportune time to erect a new, modern building that would combine under one roof the executive offices and the research and analytical laboratories. The present office building is a spacious, soundly built structure, but its white marble floors, iron grills, and golden oak woodwork seemed to them distressingly mid-Victorian. More practically, it was clear that a more efficient office lay-out might be arranged and the building of modern, centralized laboratories was fully justified. Finally, profits were good, and the excess-profits tax positively incited any reasonable capital investment.

Accordingly, they had plans drawn for a $250,000 building, and full of enthusiastic confidence laid them on Mr. Mallinckrodt's desk. He hardly glanced at them, but began plying questions. Attentively he listened to all of their answers and their arguments.

"Before we consider these plans," he said, "there is one thing I should like to know."

"What is that?" asked Arthur Boylston, who was acting as spokesman, and who confesses now that at this point he thought the new building was assured.

"This building requires an investment of a quarter of a million dollars. Just how much additional business must we do, year in and year out, to carry the overhead of interest and taxes?"

He glanced around the group, and their faces told their answers.

"And do not forget," he added smiling, "that this war is not going to last forever. Some day we shall have to take a loss on a lot of these high-cost products."

In almost identical words he reiterated the gist of his business experience: that in good times, or bad, the permanent policy is vastly more important than any temporizing program. Habitually he focused every business problem in long range until forward-looking became instinctive. To this mental habit he owed high reputation as a wise industrial leader, and in his later years, as a shrewd investor in real estate.

But those anecdotes reveal other conspicuous traits of Edward Mallinckrodt. The simplicity of his tastes was exceptional. He was so truly modest that anything even suggesting showy display was repugnant. Accordingly, he was not impressed with the thought that so important a company as the Mallinckrodt Chemical Works should have impressive offices. Because it would mean greater comfort and more efficiency, air conditioning, we may be sure, would have appealed to him strongly; but we can be equally certain that brilliantly panelled walls and chrome furniture would have no place in his scheme of decorations. When he bought, or when he built, he sought out true values in lasting standards, for as he said, "It is curious how often a good bargain turns out badly."

His keen sense of values, combined with his generosity, made his many great philanthropies extremely important. The same warm humanity that forbade any cutting of the payrolls during depressions, led him to spend both his time and a large portion of his fortune in alleviating human suffering. Chief beneficiaries of this interest were two St. Louis institutions, the Medical School of Washington University and St. Luke's Hospital.

In 1910, largely at Mr. Mallinckrodt's instigation, the Medical School was reorganized, and he was one of three who provided endowments to put upon a full-time basis the clinical departments of medicine and surgery. Co-operating with the General Educational Board, he made possible a permanent endowment of $1,000,000 for the department of pediatrics, which was officially named after him, and he personally endowed a chair of pathology. Ten years later he gave funds for the establishment of a school of pharmacy, and he and members of his family made their largest contribution in 1927 for building and equipping the Institute of Radiology. These gifts itemize but a few of his more important donations to the Medical School. From 1902 until his death, he was a member of the Corporation of Washington University, and he made many anonymous gifts to the Salary Fund, to the Woman's Building, and to promising students in fellowships or salaries.

Edward Mallinckrodt's connection with St. Luke's Hospital was as long and close as with the Medical School. The same year (1902) he was elected to the hospital board and from 1920 until his death served as its president. Unknown to his closest associates, he for many years made up the deficit and on three occasions kept the doors of the hospital open and its services unimpaired. Every Sunday, when in St. Louis, he spent several hours in the hospital, and as his co-worker, F. V. Hammar has testified, "no detail of policy, management, expenditure, or ethical hospital practice was too small to receive his scrutiny, or too large to receive his broadminded and understanding judgment."

In the same groove of his interest were his generous endowment of the Children's Hospital, (St. Louis) as a memorial to his wife who had been deeply interested in the work with crippled children, and his gift of the Mallinckrodt Chemical Laboratory to Harvard University, the alma mater of his only son, Edward Mallinckrodt, Jr.

Shortly after his reluctant announcement of the gift

to Harvard, a business associate came to Mr. Mallinckrodt with an earnest plea for funds for a small mid-west college. After careful consideration, he told his friend that this school was so small that though a student might get a first-rate teacher (say) in chemistry, nevertheless he would certainly get a fourth-rate laboratory, whereas if he went to a larger school, he would have not only one good teacher but several, and also a good laboratory. "It is better," he concluded, "to give one boy a really good chemical training, than four boys a third-rate education." Thus his philanthropies were guided by his sense of values and his long-range point of view, and in his public services, too, another personal characteristic was prominent. Mr. Hammar has testified to his close attention to detail and to his unwillingness to accept a purely titular honor. If he accepted a directorship, he must help direct: if he was president, he would soon naturally lead.

Thus also, in the conduct of his business—in his later years his good friend, Robert Brookings, urged the wisdom of throwing onto younger shoulders the responsibilities of active management in order that he might have more leisure to devote to his philanthropic and educational work, to his hobbies of gardening and art collecting, and to travel and recreation. Mr. Mallinckrodt did try sincerely to follow his wholesome but impractical advice. But he could not lay down the reins. He could never be contented detached from the business which had become his very life. In reviewing that strenuous, successful life it is clear why this should be inevitably so in the case of a man of his energy and determination. Having built a great business in the face of obstacles before which most others would have quailed, old habits could not be broken.

Although Edward Mallinckrodt was, in the material sense, a self-made man, he was spiritually and mentally the scion of distinguished ancestry. Since before 1600 the Mallinckrodt family had lived in Dortmund, Germany, and the annals of that ancient city record a long line of councillors, burgomasters, and judges of their name. The family fortunes, which were more than comfortable, had been founded in the cloth trade and were destroyed during the Napoleonic Wars. Dortmund was successively occupied by Russian and French troops, and Napoleon's brother, Jerome Bonaparte was created King of Westphalia. His misrule became so intolerable that the citizens of Dortmund sent an emissary to Paris to plead with the Emperor for justice. The emissary was Arnold Mallinckrodt, grandfather of Edward, the fine chemical manufacturer of St. Louis.

This Arnold Mallinckrodt was a brilliant intellectual and ardent liberalist, extremely poor qualifications for worldly success in the troubled times during which he lived. A Master of Laws at twenty, a Privy Councillor before thirty, he founded the *Westfalische Anzeiger*, the first newspaper in the duchy. In its pages the publisher exposed official graft and denounced princely tyranny. His memory is so cherished in Dortmund that the centenary of his death was commemorated with elaborate ceremonies and a public school named after him. But during his life, however, official recognition of his services was quite different. The French invaders sacked his printing plant. After the restoration of peace, the Prussians suppressed his newspaper for its bold support of the peasants against the great landowners. He was forced to move and become a professor in the University of Jena, a post he resigned because of a throat trouble that interfered with his lecturing. He next bought an estate at Soest. The land was badly run down and he was not a practical farmer, nevertheless he persevered until his death in 1825.

Arnold Mallinckrodt left a widow and two sons, Edward in Dortmund and Emil on the farm at Soest. The mother died two years later, and three years after her death Emil, despairing to make the unwieldy, unproductive estate profitable, sold it. Fired by the glowing account of rich opportunities in the lower Mississippi Valley in Gottfried Duden's famous "Report," he borrowed money from his cousin Gustav of Cologne, and with another cousin Julius, sailed for America. They landed at Havre de Grace and immediately set off in a coasting vessel for New Orleans. On the Missouri River, they settled at Augusta, and were joined by six other Mallinckrodt cousins.

Here the little colony of kinsmen set industriously to work, and here too, young Emil met and fell in love with Eleanor Didier Luckie, the sixteen-year-old daughter of a widow who, after the death of her husband, had brought her little family to Augusta to be near her two brothers. The young couple were married February 10, 1833. They became the parents of Edward Mallinckrodt and his two brothers who were later to found the Mallinckrodt Chemical Works.

Shortly after his marriage, Emil Mallinckrodt sold his share in the Augusta colony, and with his cousin Herman, opened a general store in Louisiana, Pike County, Missouri. Five years later he revisited Germany to straighten out complications in the family business and in the hope that a sea voyage would benefit his wife's health. While there he visited orchards and the famous vineyards of the Rhine country, and he brought back six hundred picked trees and vines with the intention of establishing a fruit farm nearby to the fast-growing city of St. Louis.

When he reached St. Louis, April 14, 1840, his trees were already sprouting and must be planted promptly; but he was resolved to set them out only on land he owned. He found a tract of thirty-nine acres to his liking between the Bellefontaine Road and the Mississippi River, which he promptly contracted to buy and upon which he set out his orchards and vineyard. That autumn he wrote to his family in Germany, "After seven stormy years the sun is emerging." He had raised good crops of wheat, potatoes, sweet potatoes and vegetables, all of which had found a ready, profitable sale. His trees and vines were flourishing. He had built a snug brick home—the house with additions and alterations stands still on the corner of Ninth and Mallinckrodt streets. And greatest happiness of all—

on September 27 his wife was successfully delivered of their first living child, a lusty son, promptly christened Gustav after the Cologne cousin who "with the generosity of father" had financially backed his American ventures. Two years later a daughter, Wilhelmine Dorothea, was born; followed on January 21, 1845, by Edward, named after the brother in Dortmund, and finally in 1847 by another son, Otto.

Meanwhile Emil Mallinckrodt had prospered. His grapes were "as thick as peas" and his peaches "often weighed two to the pound" as he proudly reported to his relatives; and his produce commanded fancy prices. He won prizes for his fruits and wines. He added a second story to his brick house. He even installed a bathroom which was a curiosity that drew visitors for miles around. He purchased a negro woman, for $450, to help his wife in the housework. He built a saw mill and a starch factory on his property. Yet for all his happiness and prosperity he seriously contemplated returning to a vineyard on the Rhine. This thought was prompted by the severe cold of the winters and the excessive summer heat which proved too much for his wife's delicate health and which he felt were bad for his growing children. During the winter of 1848-49 a devastating epidemic of the cholera broke out in St. Louis, and though the family escaped with their lives, all were more or less seriously stricken, so that he resolved to go to Germany in the spring. For five years they remained abroad, living with his brother Edward in Dortmund, but staying also with cousins in Cologne and visiting Switzerland and the French Riviera. October, 1855, saw them back in St. Louis, situated on a new farm on the Bonhomme Road, where it was hoped the higher ground would be more healthful. Here, save for the interlude of another visit to Germany, Emil Mallinckrodt lived until his death in 1892.

Young Edward Mallinckrodt had his first formal schooling while in Dortmund under one Herr Troppmann, and upon the return to St. Louis was placed, with his younger brother Otto, in the district school presided over by an Englishman who administered—the verb may be taken quite literally—a thorough training, especially in mathematics. Later the boys went to Webster College, at Webster Groves, where they boarded, returning each month for one week-end at home. Three years after the return to St. Louis their mother died of tuberculosis, and two years later their father proposed marriage by mail to Fraulein Vollmann, who with her sister kept school in Anholt, and accepting she came to America and they were married. Two more sons, Emil and Oscar, and a daughter, Adele, were added to the happy and always closely knit family. That their aging relatives in Germany might see these younger children, part of the family crossed the ocean again in the spring of 1862.

Now Edward Mallinckrodt emerges from childhood into a vivid and capable personality. He was eighteen years old. His elder brother Gustav was in Minneapolis serving as adjutant to Paymaster Finney. Otto was still at school in St. Louis. Care of the schoolboy and management of the farm were put in his hands.

Beyond these grave responsibilities he faced unusual difficulties. The country was in the throes of the Civil War and Missouri was hotly contested territory. Raids and robberies occurred frequently, and the Mallinckrodts were devoted Unionists in a region where Confederate sympathies were rampant. His position was one of personal danger, and the business affairs he was called upon to administer were made highly insecure by the scarcity of labor, by uncertain markets, by widely fluctuating prices, and even by the instability of the currency. That he operated the farm the first year with a net profit of some $1500 is high testimony to his industry, his courage, and his good judgment.

It is at this time, too, that his solicitude for chemical matters awoke. In the first letter to his family after their departure he wrote: "As to chemistry, tell Minna that I have not looked at my books since she left. I have no time to study and besides my thoughts are too much taken up with other things." We can well believe that, yet, when winter came with its less strenuous farm work, he returned to his chemical studies, and on December 27, 1863 he wrote:

I will soon be through with my present books on chemistry and shall then need some larger works. When you return, I would like very much to have you bring me some apparatus along and will send you a list in some of my future letters; such things are so much cheaper in Germany that there is great profit in importing them.

His interest in chemistry had been first roused by a set of Liebig's works in his father's library; and remembering his own interest in the natural sciences, the father had wisely encouraged the son by setting up a laboratory for him in an old brick outbuilding. Edward was deep in these studies when his older brother returned to St. Louis, having purchased his release from army service, to take his old position with the Richardson Drug Company, one of the oldest jobbing houses in the Middle West. Greatly impressed with Edward's hard, lonely life and the manful way he had grappled with difficulties, Gustav wrote their father urging that a deserved reward and a valuable experience be given to him in the form of a college course in chemistry in Germany:

As Edd has a decided taste for chemistry and natural sciences, I think also a very comprehensive brain, it would certainly be advisable to educate him thoroughly as a chemist, he then would have a treasure of knowledge of immensely great value which would enable him to go into some manufacturing business of which we have all talked about so much. St. Louis promises many advantages for that branch of industry as do also the vast mineral resources of the State. Should Edd, Otto, and I embark in such a business at some future day, I have no doubt that we should meet with success.

Here is the germ that was to grow into a great chemical business, for Edward, in turn, persuaded his father to give the youngest brother a similar training. Accordingly, in the spring of 1864, the two young men sailed for Bremen. In the autumn they entered the Wiesbaden Agricultural Institution to study under Karl Remigius Fresenius, student and later assistant of the great Liebig and himself known as "the father of analytical chemistry." They entered into student activities and spent brief vacations visiting relatives or sight-seeing, but for two years most of their waking hours were diligently devoted to the study of chemical science. They then decided to temper theory with practice—the germ of the Mallinckrodt Chemical Works was beginning to sprout—and Otto went to work in the chemical plant at Oeynhausen, while through the assistance of his brother-in-law, Edward was apprenticed to the de Haen Chemical Works, at List, near Hanover. The agreement is a curious commentary upon the changed status of the young chemical graduate:

1. Mr. M. obligates himself to be active in my factory for a period of 1½ years and during this time to submit himself to the arrangements which the rules of the concern prescribe and to which the other chemists permanently employed by me are subject.

2. In consideration of the fixing of an adequate contractual penalty guaranteed by you, Mr. M. promises to establish in Europe, within 5 years after his going from here no similar factory, nor to utilize the experiences gained here directly in any other establishment here.

3. As compensation Mr. M. pays 500 Thalers, of which 400 Thalers go to the credit of the business and 100 Thalers go to the sickness fund of the factory.

4. I, on the other hand, obligate myself, as soon as I shall convince myself that I am dealing with a young man who has the interest of the business at heart, to impart to him without reservation all information which can be of use to him.

Signed: Dr. E. deHaen

After a year and a half of hard factory work the two brothers with the sister Adele, who had been left in boarding school, returned to America. Otto went to work for the Pennsylvania Salt Manufacturing Company. Gustav remained with the Richardson Drug Company, where he had by now worked up to a department managership. But Edward, then twenty-two years old, tackled the job of building a factory consisting of "Building A," of stone, one story high, eighty by forty feet, housing the office, packing room, laboratory, drying rooms, with a lean-to shed containing a boiler and small engine for running pot mills; an acid house, twenty by twenty-five feet, with a furnace for firing eight 5 gallon glass distilling retorts which might be replaced by iron fusion pots; and a rough shed enclosing a still for the preparation of aqua ammonia from sal ammoniac and lime. The working capital was $10,000, raised by their father by a mortgage on his land.

By mid-winter the factory was ready for operation and the three brothers now joined forces. The firm of G. Mallinckrodt & Company was launched. Gustav was general administrator and sales manager. Edward was in charge of the factory. Otto was analyst, purchasing agent, and shipping clerk. The force consisted of three men and a boy. Their products were in the fine chemical field; in fact, many of the early operations were the purification and standardization of crude materials. They made ammonia, spirits of nitrous ether, chloroform (from alcohol and bleaching powder), potassium bromide and iodide.

These products brought them in direct competition with old firms in Philadelphia and New York that were strongly entrenched in the conservative medicinal field. They were unknown. Their capital was limited. On two different occasions they were almost forced to close up. Three factors staved off failure. In the first place, they established the highest possible standards for their products and their packages, which they scrupulously maintained. Secondly, their location in the fast expanding Middle West gave them a distinct time-and-transportation advantage over their seaboard rivals. Finally, but most important of all, the three brothers were well trained, courageous, intelligent, and frugal; and they all worked with might and main from sun-up till late in the night. Literally they lived with their business, having rented a house across the street from their plant; and there was no task from firing the boiler or pasting labels to improvising a new process or interviewing an important customer that they could not, and would not, successfully put through.

Success came slowly but steadily. By 1870 Building A had three stories; the lean-to engine room and ammonia shed had become permanent structures; a new building for the manufacture of nitrous ether had been completed. Thereafter hardly a year passed without some important plant addition, so that the end of the alphabet was reached in 1896. Three years after they started the force had grown to nine, including two boys and three girls. A year later the staff had doubled and by 1877, totalled over forty. Today there are a thousand persons on the Mallinckrodt payroll.

Between June 1876 and June 1877 three momentous events came into the life of Edward Mallinckrodt. On June 7, 1876, he married Jennie Anderson, daughter of a neighbor whose elder sister had already married Shepard Barclay. Theirs was a most happy, companionable marriage, and Mrs. Mallinckrodt exerted a profound, if subtle, influence upon the natural reticence and seriousness which even as a boy had been noted in her husband. On December 29, 1876, his brother Otto died suddenly and Gustav, who had been ill a couple of years, on June 5, 1877. Upon Edward Mallinckrodt fell the full responsibilities of the growing business.

Two men joined the organization who were long Mr. Mallinckrodt's mainstays—Henry W. Huning was employed in 1868 and died a vice-president after sixty-two years of service in 1931. O. L. Biebinger joined the staff in 1888; was first secretary and eventually president of the company. After 1905, four young Harvard graduates, associates of Edward Mallinckrodt, Jr., each destined to important posts, were added to the organization—Arthur C. Boylston, now vice-presi-

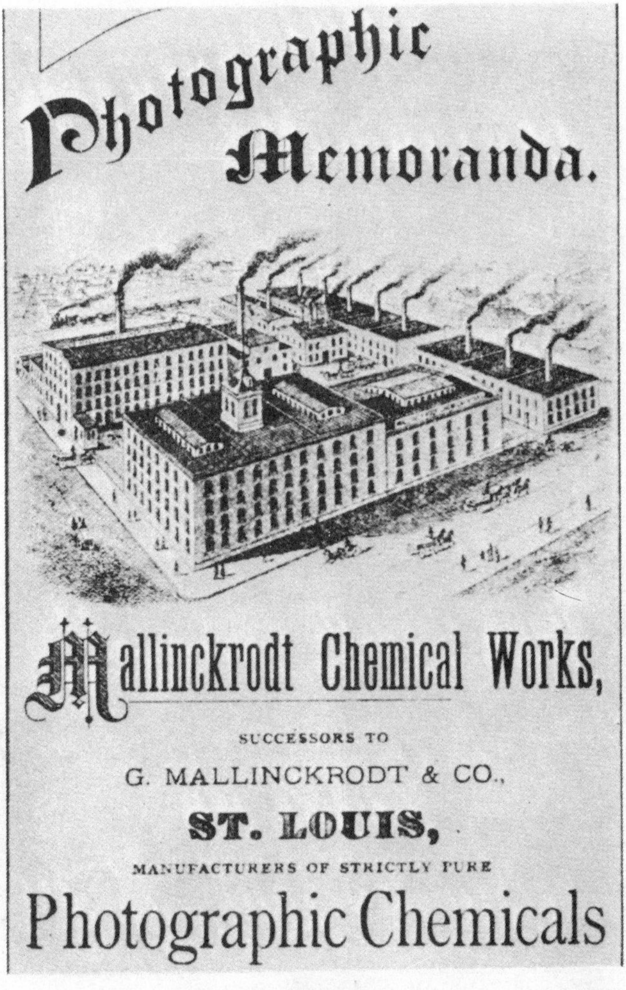

Cover of first "Photographic Memoranda" or photographic handbook used by the Mallinckrodt Chemical Works, 1882.

Oscar Biebinger

Edward Mallinckrodt, Jr.

Arthur C. Boylston

F. W. Russe

dent; Frederick W. Russe, now secretary; Wilfred N. Stull, and Harold W. Simpkins, who at the time of their deaths in 1938 and 1935 were respectively vice-president in charge of production and treasurer of the company.

In his later years Mr. Mallinckrodt depreciated his purely chemical attainments saying, "Of course, I can't claim to be a chemist any more." Nevertheless his careful training and retentive memory and the many years of daily contact with chemical manufacturing, enabled him, long after his son, who had retired from Harvard University in 1903, had taken charge of research and production, to discuss intelligently technical problems. He was truly a chemical executive, in the full meaning of both words; yet his commercial and financial abilities overshadowed his technical attainments. His sound judgment and habit of looking forward enabled him to seize upon opportunities in new chemical fields. He sensed accurately the future of photography; and though the roll film was invented only in 1884, two years earlier when the first portable camera was on the market, Mallinckrodt was making photographic chemicals. From the first the company produced bromides, and as early as 1873, he visited the salt works along the Ohio River with Herman Lerner and contracted directly for "bitter liquor," then the source of crude bromine. When Herbert Dow perfected his electrolytic process for the extraction of bromine from salt brine, he went to Mr. Mallinckrodt as the dominant maker of bromides, and offered him a substantial share in the process. The operation was revolutionary, and Mr. Mallinckrodt admitted later that this was one of the great opportunities he had failed to embrace.

Another great and revolutionary chemical opportunity, anhydrous ammonia for refrigeration, he did grasp. Applying the principles of refrigeration by liquid ammonia discovered in 1876 by Professor Lynde, E. H. Larkin, of Larkin & Scheffer, St. Louis, perfected in 1897 a shipping container, and his firm began the manufacture of ammonia for refrigeration. Within two years the Mallinckrodt works were producing anhydrous ammonia and within ten years were the largest American maker. The most serious problem was distribution, involving the continuous servicing of refrigerating plants, and this was solved in 1889 by the organization of the National Ammonia Company, a consolidation of the five leading producers: Mallinckrodt and Larkin & Scheffer of St. Louis, de la Vergne Refrigerating Machine Company, New York, Theodore J. Goldschmid, Philadelphia, and the Delaware Chemical Company, Wilmington. By the outbreak of the World War the National Ammonia Company, of which Mr. Mallinckrodt from its organization had been president and general manager, covered this country, Canada, the West Indies, part of South America, and Australia; but the synthetic ammonia process, developed during the war, changed the situation. The company undertook synthetic production in 1924, and at the same time entered into a manufacturing agreement on ammonia with the du Ponts. Four years later the du Ponts bought all the stock of the National Ammonia Securities Company, and the final piece of business done by Edward Mallinckrodt was the signing of these transfer papers with his old friend and close associate in the Ammonia Company, Colonel John C. Atwood.

At the time he was in his last illness. Although he continued active to the end, his health had been failing during the three previous years. In fact, his closest friends knew truly that since his wife's death in 1913 he had been, as one put it, "upon the downward slope." The sudden blow of Mrs. Mallinckrodt's death from an unexpected heart attack struck him deeply, and though he manfully tried to cure his deep sorrow by hard work, which he believed was the best antidote, nevertheless he was thereafter a changed man. He did in later years relax somewhat, visiting friends and family in Germany, spending more time in his gardens both at his home in the suburbs of St. Louis and at his summer place at Old Forge in the Adirondacks. His interests had broadened in his many philanthropies, his directorship in the St. Louis Union Trust Company, and in the Mercantile Library, St. Louis, the St. Louis Art Museum, and Shaw's Botanical Gardens. Upon his eightieth birthday, January 21, 1925, a host of friends took this appropriate opportunity to show him their affection and esteem. Modest to an embarrassing degree, reticent almost to shyness, without that spontaneous charm that attracts friends easily or that brilliant fascination that commands immediate recognition, this great chemical manufacturer had won both warm friends and high honors.

Three years later, he entertained at his birthday party a small group of old friends, but a week afterwards he was stricken with pneumonia. He rallied sufficiently to conclude the ammonia business, and he died on February 1, 1928. An editorial in the St. Louis *Post-Dispatch* summed up the influence of his life:

> He established a great business, which at once reflected his genius for organization and satisfied that constructive urge which . . . is the true solution of happiness for great souls. But that was far from all Mr. Mallinckrodt did . . . His benefactions had soundness and foresight. . . . He was, indeed, a singular combination of success and culture, a phenomenon all too rare among us.

AUGUST KLIPSTEIN

1848-1926

FROM Appomattox to the Marne stretched a half century of phenomenal industrial expansion in the United States. During this period, as the demand for chemicals grew, we built up in this country a heavy chemical industry (acids and alkalies, pigments and dry colors, solvents and adhesives, and the more widely used salts, except potash, of the commoner acids) more and more capable as the years passed of meeting our domestic requirements. During this same period, new chemical developments created great gaps in our chemical supplies that of necessity could only be filled from abroad.

There was developing in Europe an entirely new industrial chemical technique. The young science of chemistry was being put to work to serve human needs, and hardly a year passed without the discovery of some new chemical; some better, cheaper way of making old chemicals; some new chemical process useful in industrial operations. The discovery of synthetic dyes from coal tar is but the most conspicuous example of these nineteenth century applications of chemistry to industry.

For a number of good reasons we could play only a minor part in this stage of chemical development. Both the technical and financial resources of our own young chemical industry did not permit it to carry on the necessary research or to experiment with untried products for which there was a small, uncertain market. Since our chemical manufacturers were fully occupied with the absorbing task of supplying the fast-growing demand for standard chemicals, the task of introducing new chemicals to American consumers fell naturally to the importers. In doing this they assumed greater risks, but they performed a highly valuable service to American chemical consumers. It was but natural that the successful among them reaped rich rewards. As our chemical manufacturers kept adding to their products, the importers were continually losing markets, and in self-defence, were forced constantly to be alert to new chemical opportunities. Thus natural competition served the chemical users, until the utter dislocation of the world chemical trade by the World War and the subsequent expansion of American chemical industry into coal tar and other organic branches, severely cut down the former importance of the importers.

In the old sense that group has all but vanished, yet the prominent position they once held and the real service they long rendered are an important chapter in American chemical annals. They were pre-eminently chemical salesmen, and the tradition of service to the consumer of chemicals which they passed on is today the cherished sales policy of many of our great companies.

Among all the chemical importers of the Old School the one who perhaps most perfectly exemplifies their most successful methods and their best practices was August Klipstein. Keenly scrutinizing every technical and commercial development for opportunities to sell more chemicals; always seeking new ways of helping his customers, he was for years a leader who set the pace in the field of imported chemicals. Inevitably, since he thought originally and acted independently, he introduced into chemical selling a number of valuable innovations. He was a pioneer in technical selling. It is said that he set up the first chemical laboratory in the United States devoted to assisting customers with the deliberate intent of increasing sales. In the early nineties he distributed a little booklet entitled: "Tanning Materials: The Origin, Properties and Uses of Quebracho, Amazona, Myrobalans, etc." which for its accurate information and readable style was years ahead of the stereotyped publicity then used to advertise chemical materials. He introduced the color card in the sale of dyestuffs to the textile industry. His house was the first, so it is claimed, to establish branch offices which carried stocks in their own warehouses for the convenience of local buyers who required prompt deliveries.

To the generation that has entered chemical industry

since the war it is difficult to convey the great prestige held by the Klipstein organization. But no old-timer forgets the tremendous influence of that name. Besides being the outstanding importer of chemical materials from Europe, the firm by the turn of the century had extended its activities to every part of the world where raw materials were produced. The international importance of the Klipstein name was such that it was an *open sesame* to trading in the cities and outposts of all five continents. The Klipstein connections in Europe are almost a full roster of famous names in foreign chemical history. As an importer of raw materials, the Klipstein firm vied with the great trading companies of the world. Connections were established in China to search out sources for tung tree oil; in India for tanning materials; in the East Indies and New Zealand for natural resins and gums, and in many other trading centers of Asia, Africa, and South America for raw materials with which to satisfy the ever-increasing demands of American chemical users. No man in chemical commerce deserves better than August Klipstein to wear the proud motto of the Prince of Wales, "Ich Dien"—"I Serve."

The emphasis, too, is well placed upon the first word of that motto. For while his aggressive, modern sales methods soon became standard practice in chemical selling, neither August Klipstein nor his contemporary rivals were of a standardized type. As men, there was no typical chemical importer. Each was a marked

August Klipstein

August Klipstein's soldier father, Georg von Klipstein, who died young, leaving a widow and a son and daughter.

individualist. But however sharply their characteristics might be distinguished, all of the group shared as executives two common traits. Each one not only knew, but he also personally tended, every detail of his business, and they worked hours that to us seem cruel and unnatural. All his life August Klipstein, whenever he was in New York, was at his desk by eight in the morning, and he seldom left before six in the evening. Until his death he kept a firm grip upon the reins of his multitudinous interests.

Born of ancient German lineage and educated in Germany, August Klipstein personified the old Germanic virtues. His was the practical philosophy that believed implicitly in hard work and pay as you go. Strict in his demands on those who worked with him, he was generous to a fault. Impatient with those who gave reasons why a thing could not be done, he often lost his temper at the negative attitude.

"Thousands of reasons why it can't be done; not one why it can!" he would exclaim in great irritation. Over and over again he found a way when others had given up disheartened. His energy was prodigious and served as a spur to the whole organization, and his courage, even in his later years, was more resilient that that of many younger men. It was a courage based not in bravado but in his ability to see ahead, to reason clearly and to penetrate to the heart of a problem quickly.

His almost prophetic business vision was based upon this unusual ability to reason forward to future results from present causes. He was, for example, one of the first to sense the coming need for tannins. Extensive and wasteful lumbering operations had seriously curtailed our native stocks of tannins during the period following the Civil War. The tanneries moved from the seaboard westward, following the vanishing forests. Concentrated, even solid, extracts of the various tan barks began to make their appearance in response to this geographic problem. The tanners were forced to import more hides from the Argentine and Australia, and so became less able to move inland.

August Klipstein sensed the solution of the problem lay in the importation of materials rich in tannins and concentrated tanning extracts—and acted accordingly. He combed the markets of the world for these materials and his pioneering in this field built for him a tremendous market among tanners. His firm was the first to introduce quebracho and myrobalans to this country. This is but one of many situations in which he demonstrated an uncanny gift for forecasting the market.

To this forward-looking habit of mind, he added another uncommon ability. He was a consummate judge of men. This talent served the House of Klipstein in two ways. His sound judgment of character made him an exceptionally astute credit manager for the firm. Based upon his instinctive personal opinion, credit was more than once extended to young men or to new companies in cases where the bare facts of their financial statement would plainly indicate the wisdom of extreme caution. His appraisal of inherent honesty was but seldom wrong, and it built many small buyers into big customers who were firm friends. But his rare good judgment of men was of even greater service to the firm in discovering, and training, and promoting to positions of responsibility the personnel of a truly famous sales staff.

August Klipstein was not himself technically trained in chemistry, and he was by no means a "hail-fellow-well-met" man. Yet he recruited, instructed, and commanded a sales staff of over seventy-five men, which included such prima donnas of the older school of salesmen as C. C. Speiden, Byrd Walker, William H. Jackson, Thomas Clexton, and A. G. Wackenreuter. To control such hard-working, hard-playing individualists without quenching their fiery enthusiasms; to bind such incorrigible independents loyally to the firm; to inspire these relentless salesmen with his broader ideas of service to the customer was a triumph.

It was the force of his character, his sense of justice, his will power, his intimate knowledge of every detail of the business, and most of all, his own warm generosity (often well hidden) that bound his staff to a man at once impatient and domineering. As one of his own "Klip boys" has said, "Old 'A. K.', as we called him, had a hard shell and a soft heart. He simply hated a shirker, and he had no patience whatever with a stupid mistake. Nobody could fool him because he knew more about the business than all of us put together. For this very reason he appreciated both our difficulties and our successes and was always ready to help in case of trouble and to reward liberally for work conspicuously well done. He was a hard boss and a good employer."

An anecdote is preserved of the early days which reveals in a flash the contradictions of August Klipstein's temperament. On the wall beside his desk hung a large day-by-day calendar which it was the office boy's duty each morning to bring up to date. A new boy once failed to fulfill this simple task, and "A. K.", calling him in, jerked his thumb at the calendar and went on with the great stack of morning mail. The boy mistook his abrupt gesture for an order to close the window. On the window sill stood a great five gallon bottle of distilled drinking water. In his haste the boy clumsily toppled this water bottle out into Pearl Street. At the sound of the crash Mr. Klipstein sprang to his feet, and taking in the situation at a glance, roared at the boy. Out of the private office sprang the youngster. At his heels came his employer. Through the accounting room they rushed; down the stairs and across the shipping department on the ground floor. In desperation the boy fled headlong down the cellar stairs and took refuge in the toilet, slamming and locking the door after him. Pounding on the panels with both fists, the head of the firm informed him in no uncertain terms of his instant discharge.

The possibility of a serious accident; the clumsiness of the boy; his seeming stupidity in mistaking a plain gesture, these for the moment had all blinded Mr.

Klipstein to the true relationship of cause and effect. Thinking it over, he properly blamed himself for having given improper instructions. Quickly he made handsome amends. The very next morning he took several hours of his valuable time to go out into the country in Harlem where this boy lived and re-engage him with an apology. And it is quite in the spirit of the times and of the firm that this same boy became in due time head bookkeeper.

August Klipstein entered business as a "commercial apprentice" with Gehe and Company, drug and chemical merchants of Dresden, Germany. He was at the time sixteen years of age, having been born in Darmstadt, June 27, 1848. His father, Georg von Klipstein, was a young Hessian officer, scion of a long line of distinguished but impecunious military men. Had this father lived, it is likely that young August would have followed the family profession; but he died young, leaving a widow with a little boy and girl and a very slender inheritance to care for and educate them. Prompted no doubt by his gentle mother's influence, young Klipstein determined to carve out a business career. As a stripling he decided to become a merchant, and his first ambition was to give his mother and sister some of the little comforts and simple pleasures that he knew had been denied them. He was throughout his life this same grateful son and devoted brother. He was a brilliant scholar in the schools, first of Darmstadt and later of Frankfort-on-Main; and towards his commercial end he was tutored in foreign languages, so that by the time he was ready for business he could speak and write not only German, but also French and English, almost like a native.

His flair for foreign languages opened the door of opportunity, for when he was but eighteen, he had the chance to join his cousin, Fritz Fink, in Paris. This cousin, who later also made a name for himself in the chemical industry, was at the time employed by the old, distinguished firm, Tollard et Cie., commission merchants, who had been carrying on an international business in seeds and grains since 1796. Fink introduced his cousin to his employer, and Paul Tollard was so impressed with young Klipstein's command of French that he immediately offered him a position with the good salary of seventy-five francs a half month. So August Klipstein became a wholesale seedsman and within the year was sent to England to represent the firm in London and Manchester. This experience proved extremely valuable, for it not only enabled him to acquire practical proficiency in the use of English; but it gave him a gruelling training in one of the most highly speculative of all the world's markets.

Returning to Paris in 1868, his knowledge of languages again made possible an opportunity which proved to be the turning point in his life since it brought August Klipstein definitely into the chemical field. The prominent firm of Edm. Renault et Cie. asked him to take charge of all their foreign correspondence. He accepted their offer, and again gained a valuable training which gave him not only a broad view of international chemical trade, but also an extremely intimate knowledge of the details of this complicated business.

His work in Paris was violently interrupted by the outbreak of the Franco-Prussian War. A trusted employee was thus unwittingly turned into an alien enemy. He left France on one of the last trains that ran between the two countries and reported to the military authorities at his home city of Darmstadt. Because of his youth and slender physique he was placed in the infantry reserve. Accordingly, he went on to Dresden where he again joined the staff of Gehe and Company with whom he had served his first apprenticeship.

After a year and a half, the war having ended, he was recalled to Paris in 1872 by Renault who promptly put him in charge of all their foreign business. He had hardly settled fairly into his new work when he was summoned one morning to the office of the head of the firm and the proposal made that he should come to America to represent them in the United States.

Here was the opportunity he had been patiently seeking and for which he had been so painstakingly training himself. He knew very well the demands of the American market for chemicals. He appreciated thoroughly the rapidity with which that market was expanding. He had at his finger tips all of the European sources of chemical supply, and he enjoyed close personal connections with two great chemical mercantile houses, Renault in France and Gehe in Germany.

Moreover, he spoke English. His forward-looking logic completed the vision of a great career in the United States for one August Klipstein, chemical merchant; and he stipulated that, if he came to America, he should represent Renault as their agent, not as their branch manager. His terms were acceptable, and he sailed from Liverpool on the S. S. "England", June 26th, 1872, landing in New York on July 10th.

He immediately opened his office on Wall Street, and almost from the first met with a success that justified his expectations. "Rome was not built in a day" was one of his favorite sayings, and he laid down very solid foundations. Always alert for new opportunities, he was scrupulously careful not to expand more rapidly than his accumulating resources could finance without undue strain, for he was also fond of quoting "Don't spread your butter too thin." Successively he moved to larger and ever larger quarters, first on Platt Street, then to 52 Cedar, and in 1890 he purchased the two buildings at 122-124 Pearl Street, which for twenty years were the Klipstein headquarters. A newspaper of 1890 described the building as follows: "The imposing structure four stories high, extending through to Water Street, is completely lighted by electricity, has electric elevators, and contains the laboratories in which the Company's experienced chemists conduct tests and scientific researches."

In 1908 this Pearl Street building was gutted by a great fire. All the records were lost and a valuable stock of dyes and chemicals destroyed. The very

A. Klipstein & Company were one of the first chemical importers to maintain spot stocks at various points: their Chicago branch in the early nineties.

Mrs. Klipstein, with her children, August Klipstein, Jr., Herbert C., and Louise A. (now Mrs. A. W. Shields).

next morning in offices lent by a business neighbor, August Klipstein gathered his staff together and carried on. Within a few months they all moved to the large eight story office and warehouse at 644-654 Greenwich Street, beside the Ninth Avenue Elevated.

Tanstuffs and dyestuffs—those two big gaps in our chemical supplies—were from the very first specialties of the House of Klipstein, but the wide development of all branches of our chemical and dye industries at the present time makes it difficult for us to realize how short a time back we were still largely dependent upon foreign makers for even the commonest of our heavy chemicals. The sales figures of this great importing house for the year 1908 reveal not only this, but also the extent to which the business had grown:

Quebracho Extract	33,000	Tons
Mangrove Bark	10,000	"
Myrobalans	11,000	"
Sumac	1,000	"
Kauri Gum	2,000	"
Bleaching Powder	8,000	"
Carbonate of Potash	2,800	"
Caustic Potash	2,200	"
Epsom Salts	2,000	"
Aniline Salt	1,500	"
Barium Chloride	1,300	"

and over a thousand tons each of such chemicals as prussiate of soda, bichromate of soda, sulfate of alumina, oxalic acid, and white arsenic.

The large sale of the various tanstuffs is significant,

for not only was the firm responsible for the first introduction into this country of quebracho and myrobalans; but Mr. Klipstein's personal interest in tannin supplies led him to purchase in Florida sixty thousand acres of mangrove tree bearing land, a vast plantation still owned by his family. Naturally, at no time, did the tonnages of the highly concentrated coal-tar dyes reach such imposing figures; but Mr. Klipstein was one of the first to recognize the merit of synthetic colors, and as early as 1880 he secured the sales agency for this country and Canada of the Society of Chemical Industry in Basle, Switzerland (then known as Bindschedler & Brusch), an agency he retained till after the World War when it was released in 1920.

Supplying primarily the multitudinous chemical needs of the leather and textile industries, the business reached out to supply chemicals for all the consuming industries. On January 1, 1894, the personal enterprise of August Klipstein became the corporation, A. Klipstein & Company. August Klipstein was president; William H. Jackson, then manager of the Philadelphia office, vice-president; and Ernest C. Klipstein, not a relative, but a partner, was the treasurer. Shortly before, in 1891, Mr. Klipstein had branched out into chemical manufacturing and organized the Bulls Ferry Chemical Company. At Edgewater, N. J., this enterprise operated one of the first plants for the manufacture of industrial chemical specialties. This company, for which the A. Klipstein Company served as selling agent, became one of the largest makers of softeners, sizes, and sulfonated oils for the textile industries; of synthetic tans, bates, and dressings for the tanners; and more recently of synthetic resins for the paint, varnish, and lacquer manufacturers. In the meantime a network of branch sales offices had been established, the points covered being Philadelphia, Boston, Chicago, Providence, R. I., Charlotte, N. C., Montreal, and Toronto.

Without outside assistance, through his own persistent zeal, August Klipstein had become the important figure in the chemical world that the young chemical merchant with a flair for languages had dreamed of as he crossed the Atlantic in the summer of 1872. He had developed an amazing variety of enthusiasms and interests. He was assuredly a living example of another of his favorite quotations: "If you want to get something done, go to a busy man." He was ever ready to help his friends and his family. He was never loath to assume a new responsibility. Strenuously he lived an interested and an interesting life. He had the clipping and memorandum habit, and his pockets were always stuffed with cuttings from newspapers and trade journals, odd sheets of paper, old envelopes and what not upon which he was constantly jotting down with a stubby pencil quotations from his voluminous readings in history, travel, philosophy, and poetry. Day or night, for he kept a pad at his bedside, he would note new business ideas, and he never failed to record the name and address of any new firm he heard of or whose place of business he passed, which seemed a likely buyer of chemicals. Off that note would go to the salesman in that territory, and he always insisted upon a special report on all of these new personal prospects of his.

His recreations were chiefly intellectual—in an easy chair with a good book in any of three languages, he could be quite oblivious to his surroundings, and he thoroughly enjoyed the opera and the theatre—but he rode horseback and took long walks. He was for years a prominent member of the Riding and Driving Club of Brooklyn, and one of his favorite stories was of crossing the span of Brooklyn Bridge on a narrow plank walk with his daughter, Louise. As a result the frail youth who was refused for military service became an extremely robust man with a great store of energy.

His tastes were Spartan in their simplicity. Ostentation of any kind was repugnant to his nature. His whole way of living was modest in comparison to most men of his means; but he insisted always on the best of everything—foods, clothing, furnishings and equipment of all kinds—as the truest economy, while his good sense dictated that all of these good things of life should be for wholesome use and enjoyment, not for display. Those who only knew him behind his formidable business manner would never suspect that among his friends and family he was a great tease and practical joker who could laugh, even when the joke was turned upon him, till the tears rolled down his cheeks; a jolly, companionable father who delighted to take his children to the Zoo and the Museum of Natural

History and who instituted as an annual family custom a great gala trip to the circus.

Business was the dominating influence in his life, a gigantic game of chess which he took a keen delight in winning. No other field of activity, no profession or pastime could offer greater opportunities for displaying his varied talents. Because his interests were international, August Klipstein was truly a man of the world. He knew intimately many of the great and near-great in every corner of the world, and the number of men who rose to prominence with his encouragement and backing is a tribute to his keen judgment and remarkable personality.

"Keep your sword bright," meaning body, spirit, and intellect, was another of the sayings August Klipstein was fond of quoting and which he practiced as well as he preached. To the very end he remained the actual head of the firm of A. Klipstein & Company and the president of the Bulls Ferry Chemical Company. Having founded the business, carried it successfully through a fire, two panics, and a war which of necessity shook loose the very foundations of the chemical trade as he had always known it, he died in his seventy-eighth year, on January 8, 1926. He kept on to die as he always wanted to die, in armor, his bright sword untarnished, broken clearly at the hilt.

ERNEST C. KLIPSTEIN

1851-1922

IF, over sixty years ago, a New York oculist had not first made the mistake—the same which was later to be repeated by many others—of confusing the families of Ernest C. Klipstein and August Klipstein, a great deal of interesting and important chemical history would never have been written. As a matter of fact, it is necessary to go back to Germany to the late years of the Seventeenth Century to find a common ancestor of these two Klipsteins who were so prominent and for many years so closely connected in the American chemical industry. They were brought together in 1875 by the merest chance.

At that time, inspired by the traditions of his family, Ernest C. Klipstein was working in a drug store in his native town of Marshall, Virginia, as the proper preliminary training for a medical career. August Klipstein, who had come from Germany a few years before, had established himself in New York City as an importer of dyestuffs and chemicals.

Ernest Klipstein's sister had broken her eyeglasses and sent them to be repaired by an oculist in New York who, in the polite Victorian language of the time, enjoyed the patronage of the rising young chemical merchant, August Klipstein. By some mistake the glasses of August Klipstein were shipped off to the sister of Ernest Klipstein in Virginia. In those days this meant a serious delay. We can well imagine that the busy and exacting Mr. Klipstein in New York took such carelessness tartly, and no doubt the thin jacket of some hapless oculist-apprentice was properly smoked. But this ill-contrived error had a most fortunate conclusion. Quite natural curiosity roused by the common surname, led to correspondence between the two families.

The letters from the chemical importer stirred the young drug clerk. He was quick to realize that a dyestuff and chemical warehouse in New York held much greater opportunity than a small town pharmacy in Fauquier County, and he was not slow to grasp the chance of improving his prospects. With no better introduction than the oculist's mistake and no stronger claim upon the consideration of his prospective employer than a common great-great-great-great grandfather in Germany a century and a half back, he set out for the Metropolis, resolved to enter the chemical business through the door at 32 Platt Street over which neat gilt lettering on a black sign proclaimed:

A. KLIPSTEIN

Importer and Commission Merchant

DYE STUFFS, TAN STUFFS, CHEMICALS, &c.

As August Klipstein confessed years later, he was engaged with some misgivings. It was agreed that he was to be paid eight dollars a week and that out of this handsome salary he should board himself. In view of the name and the custom of that time, when young assistants often slept in the store or warehouse and ate with the proprietor's family, these arrangements were unusual. Plainly, August Klipstein did not intend that even the thin thread of relationship was to be pulled by his new clerk from Virginia. On the other hand, likely enough Ernest Klipstein still looked upon the chemical business as but a stronger springboard to medicine. Whatever may have been the natural misgivings and contrary ambitions at that time, ten years later August Klipstein took Ernest Klipstein into equal partnership in the firm which at that time became known as A. Klipstein & Company.

When he first came to New York, Ernest Christian Klipstein was twenty-three years old, having been born on Christmas Eve, 1851. The homestead that was his birthplace had been built by his grandfather and is still in the family's possession. It stands close to the town of Marshall, in the northwestern corner of Fauquier County, amid rolling farmland that stretches to the bold rampart of the Blue Ridge Mountains but a few miles away. It was across the Blue Ridge, in the city of Winchester, at the head of the Shenandoah Valley, that his great grandfather, the first of the family in America had settled at the close of the Revolutionary War.

Klipstein family group, about 1898—left to right, J. H. Mills (brother-in-law), his wife, Ernest H. Klipstein, Mr. and Mrs. E. C. Klipstein, Gerald Klipstein, and fox terrier, "Rex."

E. C. KLIPSTEIN

Dr. Philip Klipstein had crossed the Atlantic in a British troopship, the surgeon of the division of Hessian mercenaries brought over to fight the rebellious colonists. After the surrender at Yorktown, the young German doctor, since he liked the country and bore no grudge to his victorious neighbors, decided to stay in the United States. He selected Winchester, the most important town of northwestern Virginia, as his future home. Trained medical men were very scarce in the country in those days and he was considered a real addition to the community. His skill as a physician, his university-trained education, his liberal ideas, his friendliness and sound character, soon won him a welcome on personal as well as professional grounds. He married and became the father of five sons.

One of these five sons, Philip Klipstein, Jr., inherited not only his father's name but also his profession. With that sturdy independence that seems always to have characterized the Klipstein men he chose to leave Winchester, and rather than inheriting his sire's practice, to make his own career in Fauquier County just east of the mountains. He married Sarah Ball, a pretty daughter of the distinguished Virginia family founded by the doughty Colonel William Ball. An uncle of the second Dr. Philip Klipstein's bride had been a Quartermaster General of the Continental Army and her aunt Mary Ball was the mother of the great George Washington himself. With such influential social connections, the young physician, who had built his home on the outskirts of Marshall, soon assumed a rather important position in aristocratic Fauquier County.

Dr. Philip and Mary Ball Klipstein had three sons, the youngest of whom, Philip Augustus became a merchant-farmer in Marshall, dividing his time between the management of the home farm, which he inherited, and a business in the town that combined a country store and produce commission brokerage. He fought through the Civil War with the Eighth Regiment of Virginia Infantry, attached to Pickett's Division, returning after hostilities to his farm and store at Marshall. Here he died in 1905. He married Amanda Louise Hixon, a daughter of James Hixon, a patriotic Quaker who, despite his religious scruples, had fought bravely in the Continental Army during the Revolution. Shortly after that war he had left his old home in New Jersey and settled in Louden County, Virginia. Like the younger Dr. Klipstein, he had married a daughter of a well known Virginia family, Mary Hampton. To his daughter Amanda and her husband, Philip Augustus Klipstein, there were born three sons and two daughters. The oldest of the sons was Ernest Christian Klipstein.

The youthful "E.C."—as years later he was known to distinguish him from his older chemical partner "A."—was just ten years old when his father went off to fight in the war between the states. Those must have been exciting times for the family. Just to the east of Marshall are Bull Run and Manassas and Fairfax Courthouse. Important battles were fought all about them. The Klipstein farm flanks the main road going west through the gap in the Blue Ridge that comes out at Strasburg in the Shenandoah Valley. For four years the armies, first in grey and then in blue, marched back and forth, back and forth over that highway. There were cavalry raids, counter attacks, and hurried retreats in bewildering, constant succession. Worst of all, marauding parties of semi-independent guerillas kept scouring through the nearby gap.

In the midst of these dangerous excitements young Klipstein spent four most impressionable years of his youth. They steeled his physical courage and tempered his tolerance; but as he later most sincerely regretted, they interfered with his schooling. He managed, however, by hook and crook to get sufficient primary education to enable him to attend Roanoke College for the term of 1867-68 and to acquit him creditably. But the family fortunes, sadly depleted by the war, forced him to leave college, and the next three years he taught school in western Virginia. The following year he pioneered as a school teacher out in what were then the wilds of Arkansas; but he returned to Marshall determined to become a physician like his grandfather and his great grandfather. With money saved from his schoolmaster's stipend, he enrolled in the College of Physicians and Surgeons in Baltimore, eking out both his capital and his experience by working during vacations in the pharmacy in his native town. It was at this point that his sister's broken eyeglasses and a letter from a namesake in New York switched him to the chemical career in which he was to win such distinguished success.

When as a very modest assistant clerk first he joined the staff of August Klipstein he observed that, in the business of importing dyes and chemicals it was going to be necessary for him, if he were to climb up the ladder, to learn both German and French. He knew not a word of either. But in the office he had plenty of opportunity to learn German, and this he assiduously set out to do, begging that his employer and his fellow employees, several of whom were Germans, talk this language to him, deliberately struggling with German trade papers and chemical textbooks, and later taking on the German correspondence of the firm. French presented a more difficult problem, which he solved by going to board with a French family. Having mastered French, since business with South American countries was beginning to open up in dyewoods and tanning extracts, he left his French friends and went to live in a Spanish household. He was a ready, natural linguist, for later having a mastery of both reading and speaking German, French, and Spanish, he acquired a good working knowledge of Italian, Dutch, Swedish, and Portuguese. When his own sons came along, he insisted upon always talking to them in German, on the theory that they would thus be assured at least two languages. German became so the family habit that the boys never spoke to their father in English.

Languages were not the only studies that the young

ex-medical student from Virginia followed through vigorously during the early days of his association with the House of Klipstein. He brushed up on the little chemistry he had been taught and went rather seriously into the complicated matters of dye applications and the uses of tanstuffs. In the meantime he worked his way out of the office onto the road and became in short order one of the best salesmen.

Young E. C. Klipstein liked to meet people. He was a thorough-going extrovert, interested in men, in materials, in processes. Friendly and helpful by disposition he was an early, living exponent of that personal service to the customer upon which the older chemical merchants so solidly built up their trade. He proved to be the ideal "contact man," while August Klipstein was possessed of remarkable financial and executive abilities which found their best and most useful expression in the general administration of the business. Together they were a strong pair, complementing each other's strong points, a truly logical partnership that built up an organization of forty salesmen, serving over two thousand customers.

It soon came about that E. C. Klipstein, having organized the sales end of the business, began more and more to take up the work through which he made one of his great contributions to the chemical development of this country. He became the contact man for the partnership with their foreign suppliers. The Klipstein business was fundamentally that of importing all sorts of chemical materials. Every other year "E.C." took long trips abroad. Regularly he visited the Society of Chemical Industry in Basle, Switzerland, for whose dyes the Klipstein house was then American agent, and Messrs. Eichoff and Fuchs, in Frankfurt, who were the firm's chief suppliers of various heavy chemicals.

These periodical calls upon the principals completed, he was accustomed to go off on one of his famous chemical scouting trips. He visited different countries, inspecting the finest plants in all the various chemical consuming industries to see what improvements they were making in the technique of using chemical processes and calling upon many chemical manufacturers to learn what new chemical products they had developed. From these scouting expeditions of his he brought back to the United States a number of successful chemical specialties. Thus he rendered notable services, especially to textile mills and tanneries, by introducing into this country many new bleaches, wetting out agents, mordants, scouring and cleaning agents, dehairing preparations, bates, fillers, and what not. He was the first to import vat dyes and synthetic tanning agents, and he introduced the use of formic acid in dyeing and of sulfonated oils in calico printing.

These tours of chemical exploration carried him even to South America and the Far East. The supply of our native chestnut bark for tanning was beginning to run out, and he introduced a number of tannin bearing materials, such as myrobalans from India and mangrove bark from Africa, and most notable of all, quebracho from the Argentine. This last proved at first a disappointment. Its exceptionally high tannic acid content was useless until it was discovered in Italy that the extract of quebracho might be rendered soluble in water by boiling it for an hour in sodium bisulfite. This solubility problem solved, quebracho rapidly increased in favor among American tanners, and for years the House of Klipstein was the exclusive selling agent in the United States of the Forestal Compagnie of Buenos Aires who enjoyed a monopoly of quebracho exports from Argentine.

One chemical importation of E. C. Klipstein was destined to have important after-effects not only upon his own business career, but also upon the course of chemical industry in this country. In Germany, about 1898, he found carbon tetrachloride. He established an exclusive sales agency for the United States, and began importing it for sale as a non-flammable dry cleaning agent. His alert brain conceived the original notion of packaging carbon tetrachloride and selling it for household use. Recognizing that its chemical name would be a distinct handicap he coined the word "Carbona" and organized the Marshall Chemical Company, with a plant in his home town, which he put in charge of his brother, James Hampton Klipstein.

In the development of a market for Carbona, E. C. Klipstein displayed to an unusual degree one of his outstanding business characteristics. Once convinced of the merit of some chemical specialty, he was willing patiently and with dogged perseverance to push it on and on to commercial success. Accordingly, he spent eight years and over a hundred thousand dollars introducing "Carbona." Eventually its coined name became a household word throughout the land and sales mounted to the millions of bottles.

Having scored this commercial triumph his keenest interest in "Carbona" waned, and since the business had grown to proportions that made it a serious rival to his more serious chemical interests, in 1909 he sold three-quarters of the common stock to Abraham Wineburgh, brother of the street car advertising agent. This proved to be a wise move, since Mr. Wineburgh, by using leftover car space sold him at a low rate by his brother, was in a position to carry forward a great publicity campaign. Klipstein continued to supply the carbon tetrachloride, and in 1914 when supplies from Germany were cut off, this connection led to an alliance with the Warner Chemical Company and shortly to the organization of the Warner-Klipstein Company. These new enterprises were a turning point in E. C. Klipstein's business career from chemical merchandising to chemical manufacturing.

When the war broke out, the partners of the house of Klipstein were not in agreement as to the policy they were to pursue. August expected a repetition of the Franco-Prussian war and looked for a quick German victory. He was, therefore, inclined to tide over in any way the dearth of supplies of imported chemicals and to make the most of the opportunities which their intimate, expert knowledge of all chemical markets gave them in the wild scramble for all kinds of chemical

E. C. Klipstein in the offices of his German agents, Eichoff and Fuchs, of Frankfort. Left to right, Emil Fuchs, E. C. Klipstein, Max Eichoff.

Mr. Klipstein's home at 116 Prospect Street, East Orange, N. J., taken in 1898, with the two older Klipstein boys, Ernest and Gerald, and their next door playmate, Marion Spaulding.

supplies. Ernest Klipstein believed that the struggle would be long drawn out and foresaw that many changes in international chemical trade would be the inevitable outcome. Accordingly, he was anxious to undertake the production for themselves of certain chemicals and dyes. As sensible partners they agreed to disagree amicably. While August did not care to invest any of his own or the firm's capital in manufacturing ventures that he felt would be short-lived, nevertheless he raised no objection to Ernest doing so upon his own initiative and at his own risk.

The result was that in 1915 the firm of E. C. Klipstein & Sons began operations in a new factory building at Carteret, N. J., producing sulfur black. The "& Sons" were Mr. Klipstein's two older boys, Gerald and Ernest, their younger brother, Kenneth being still in school. Gerald soon went overseas with the American Expeditionary Forces, and Ernest, who previously had been chemist for the Dan River Cotton Mills down in Virginia, stayed on to run the new chemical plant. It expanded rapidly.

A neighbor at Carteret was the Warner Chemical Company, producers of phosphates and of carbon tetrachloride, which through "Carbona" brought Dr. Lucien Warner and E. C. Klipstein together. Warner needed chlorine badly. Klipstein was proposing to make anthraquinone, requiring chlorine for the aluminum chloride necessary for the Friedel-Crafts reaction, and he needed more caustic for the growing sulfur black output. Upon these common needs they decided to build an electrolytic alkali plant. They took in as a third partner, H. R. Nelson, whose electrolytic cells had been developed in the Warner plant. They selected South Charleston, West Virginia, being close to both brine and coal, as an ideal location, and they built the plant of the Warner-Klipstein Company.

Right next door E. C. Klipstein & Sons erected a new and larger dye plant, and began extending their line by adding anthraquinone, hydron blue, sulfur blue, and later a rather extensive line of coal-tar colors and intermediates. With a war-made famine in dyes neither greater output nor new items could be developed fast enough to keep pace with the demand. At the same time a greedy market for caustic, at prices ranging from six to eight cents a pound, made the Warner-Klipstein operation a humming success. Throughout the war expansion was rapid at both plants. The war ended: contraction and reorganization were necessary.

Mr. Klipstein promptly shut the Carteret plant and moved the sulfur black unit to South Charleston. A conscientious research was undertaken to cut the costs and raise the quality of the line of dyes and intermediates. He sold his common stock in the Warner-Klipstein electrolytic enterprise, and that company was reorganized into Westvaco Chlorine Products Corporation, which in turn absorbed the older Warner Chemical Company.

E. C. Klipstein had become a whole-hearted chemical manufacturer. Chemical trading interested him no longer, and in 1922, though still retaining the joint office of secretary-treasurer, which he had held for so many years, in his old partnership of A. Klipstein & Company, he sold most of his stock to his associates. In like manner, he continued to serve as vice-president and director of the new Westvaco organization, in which he had still a considerable interest, of the Carbona Products Company, and of the Bull's Ferry Chemical Company and the Manetto Company, both A. Klipstein & Company subsidiaries.

He now concentrated his energies and his experience in up-building the dye manufacturing operations of E. C. Klipstein & Sons. Though he lived but a short year, nevertheless he laid down so firm a foundation in this new and untried field of activity that for ten years his sons continued to build this business till in 1933 it was merged with the Calco Chemical Company, subsidiary of the American Cyanamid Company.

After but a brief illness, he died, on Sunday, April 29th, 1922. At the time, one of his oldest and keenest competitors, Herman Metz said, "E. C. Klipstein was one of the most energetic and most lovable men who was ever in the chemical business; and that is a mighty rare combination of good qualities." It is a shrewd bit of characterization.

Mr. Klipstein was never idle. Yet he always found time for his friends and his family. He had married in 1888, Grace Lilian Mills, born in Hamilton, Ontario, a granddaughter of Samuel Mills, who upon the foundation of the Dominion of Canada had been one of the original Senators appointed by the British Crown.

His three sons and his business were undoubtedly the always dominating interests of his life. But he always had a hobby. As a younger man he was interested in photography. About 1898 he took up golf, and for many years he was one of the most faithful players at the Essex County and Baltusrol clubs. For his age he played a capital game, and his sons attribute their prowess to his careful training which began when each reached the age of eight. In 1916 an accident at the Carteret plant cost him two of the fingers of his right hand, which so badly interfered with his golf grip, that in his later years he played less and less.

But more and more he devoted himself to experimental work in the laboratory. Research became his recreation. At three score years and ten he plunged into the turmoil of dye making in the midst of the hurricane conditions of the World War. Having spent a full life in the service of the American chemical consumer, introducing new materials and new processes, at its very close, he became a pioneer dye manufacturer, helping manfully in the establishment of this branch of chemical industry in the United States.

MARTIN DENNIS

1851-1916

IN the dark depths of the depression period, following the panic of 1873, a retiring, serious-minded young man learned with dismay that he must abandon his studies and seek immediate employment. Most regretfully his father explained that he was no longer financially able to forward his cherished ambitions. This was a heavy blow. For young Martin Dennis, with a purpose remarkably fixed for one so youthful, had set his heart upon becoming a surgeon.

He was admirably equipped for such a career. Not only had he been a prize scholar at the famous old Newark Academy, but he had won honors as an undergraduate at Princeton. To his studious bent was added a keen interest in science, especially in chemistry and biology. These interests eventually brought him into the chemical industry and found further expression in his life-long hobbies. As a mere boy he had learned to identify all the birds of his neighborhood. While at the University he and his friends had tramped over the richly mineralized hills of northern New Jersey on a geological expedition that laid the foundations of his great collection of mineral specimens. Many years later, after he had fought a hard-won industrial battle to success, he spent much of his time at Bonniebrook Farm out in Sussex County, New Jersey, where he raised blooded stock of various kinds and engaged in scientific forestry with the idea of demonstrating a method of profitably utilizing the large areas of stone-ridged hills of that section of his native state.

With such tastes and training his choice of the surgeon's profession was far deeper than a boyish whim. The prospect of a sudden, forced entry into the hurly-burly of business was not pleasing. An opportunity to enter the urban street railway business, then operating horse car lines, he declined on account of his distaste for the almost constant bickerings between the operating company and the city fathers who had granted the franchises. The second opportunity came in the very new field of electricity for lighting. Although this new departure was in many ways alluring, the young man noted a great reluctance on the part of storekeepers and home owners to change over from gas lighting then so generally in use. Accordingly, he decided not to join up with this new industry. With all the wisdom of hindsight, it may be justly said that young Martin Dennis lacked vision, but in his judgment he felt it better to seek employment in some much longer established enterprise. This chance came in the tanning industry.

Two of his uncles, David H. McAlpin and George S. Rose, were operating a large fancy leather plant at Yonkers, New York, producing bookbinding and hat sweat leathers, glove leather, and a wide range of commercial leathers of many weights and finishes for the manufacture of novelties of all descriptions. The business had grown to a point where it needed the supervision of an energetic and alert manager with a chemical training. This opportunity caught the imagination of the young man, for he realized that the making of leather was one of the oldest and most complex manufacturing activities known to man. He quickly accepted an offer made to him by his uncles to superintend their plant.

Descended from Colonial and Quaker stock, that had settled in Newark, New Jersey, when Robert Treat had first come there from Milford, Connecticut, to make a new settlement, Martin Dennis was born in that city on February 8th, 1851. Aside from the routine of formal education his boyhood life was filled with interesting developments. From his mother he early learned the lasting value of a truly Christian character. From his father he gained a deep interest in life from many angles, including agriculture, animal husbandry, botany, horticulture, and chemistry. His preparatory school, the old Newark Academy, was then a military school and he became in his senior year the captain of cadets. Music was one of his great delights. He had an unusually fine baritone voice and played well on the piano, flute, and guitar. At Princeton, where he was graduated with the Class of 1873, he organized the first Glee Club and devoted his studies, as much as possible in the more rigid curriculum of that day, along scientific lines with particular attention to chemistry and biology.

In 1878 he married Carrie Cooper Rose of Brooklyn, New York. Of their four children, one died in infancy and another daughter, Mrs. Gayle L. Young of Newark, New Jersey, in 1922. His oldest child, Mrs. Harris F. Brownlee, resides in Danbury, Connecticut, and his only son, Harold Dennis, who now heads the chemical company he founded, lives in Newark, N. J.

Devotion to his family and deep interest in his business were his chief concerns. However, he possessed an inquiring mind and a remarkably retentive memory so that his constant reading on historical and scientific subjects supplied him a broad range of interesting and valuable knowledge. In the last years of his life a serious heart condition compelled him to slacken his energetic activities, but he philosophically and cheerfully carried on until his death in 1916.

From the day he undertook the management of the factory of Rose, McAlpin & Co. at Yonkers, he applied the principle of strict attention to the usually undervalued details of manufacture. The care given to these small matters soon showed results. A more uniform output and consequent better grading of the daily production was one of his first triumphs. The business grew rapidly until it employed 600 hands.

During these years young Dennis developed several

Martin Dennis

A tintype of Martin Dennis, left, and Harry E. Richards, taken at the time of their graduation from Princeton, 1873.

new leathers, the most noteworthy being the manufacture for the first time in this country of types of glove leather which had been until then imported from Europe. To do this required painstaking research which he personally directed. These research problems led him into the detailed study of ancient, present, and theoretical methods of tanning and finishing hides and skins.

Before our cavemen ancestors could chip a flint arrowpoint, long before they had molded and baked the first crude clay pot, probably even before they had learned to make and control fire, they had used the skins ripped off the great beasts they had slain with clubs and stones in defence or for food. This was an easy, simple discovery, since the advantages of the fur for comfort and warmth were obvious. But these raw hides quickly rotted, becoming putrid and useless, and it took real inventive skill to devise ways of keeping them and yet preserving their original soft pliability. Doubtless, these first discoveries were happy accidents. Hides dragged through wood ashes or some clayey earth kept longer. Thus, alkalies entered tanning. A hide that soaked in some stagnant pool, thick with decaying vegetation, was found to be miraculously preserved. In this manner the first tannins were employed.

It was a very real genius, however, who grasped the meaning of such accidents, and having unravelled the secret of cause and effect, began deliberately to smear and soak the skins he was preparing for human use. With justice that first tanner may lay claim to having been our first chemical process manufacturer. For though he did so all unwittingly, nevertheless he was putting alkalies and tannins to work to accomplish chemical results he did not understand. To this day much of the chemistry of tanning is still an obscure, unsolved mystery, but throughout the ages, by a long series of such lucky accidents and fortunate cut-and-try practical experiments, the tanner's complicated art was perfected. Tanning practice thus became quite fixed by rule of thumb, and tanners—as Martin Dennis was soon to discover—were very loath to make changes in their long-accepted formulas and methods. Nor is it quite fair to charge them with stubborn conservatism. Tanning is a long process, involved with several distinct operations, the wherefores of which are well known but the whys of which are mysterious. Accordingly change at any point may often have obscure, but definite effects upon the quality of their finished goods. Having once worked out a process that yielded satisfactory results, it is hardly blameworthy to be cautious in making changes that might upset a year's work and spoil much valuable leather. Since till recently the chemist was wholly unable to demonstrate reasons, it was plainly good policy to let someone else do the experimenting.

Martin Dennis was one of the first to provide a logical basis for changing this ancient habit of rule of thumb tanning for he was one of the first to provide a basis of chemical logic for his innovations. The fascinating story of leather making charmed him and he was captivated by the chemical mysteries of tanning, so with youthful enthusiasm he delved into the history and technique of the industry. Thus, he came across the early ideas and attempts to use the mineral kingdom, as against the vegetable kingdom, as the source of an entirely new type of tanning compound to produce a quality of leather possessing many desirable characteristics never before obtained from any of the centuries-old methods of tanning.

The salts of chromium showed the most promise of success. Augustus Schultz had been the one who had persevered the farthest along the road in the use of chromium. In fact, he had been issued a patent which called for treating properly prepared hide matter in a bath containing an acidulated sodium bichromate solution until the fibre was well saturated and then in a second bath containing a solution of some suitable reducing agent for the bichromate. The result was to fasten the chromium to the fibre of the hide or skin, thus rendering it no longer susceptible to putrefaction, which is the chief function of a tanning compound.

To Dennis this seemed to be an unwise approach to the problem. It involved the use of a rather violent chemical action directly in contact with the delicate fibrous structure of the hide or skin. He felt it would be far better to prepare a completed chromium compound having tanning properties into which hides or skins could then be placed and the tanning action carried out in a simple, safe, and direct one-bath method.

With this in mind he went at the problem. As may be imagined many early attempts failed. Gradually moderate results were achieved and by careful procedure a chromium chloride liquor was developed which, time after time, tanned various types of hides and skins. He applied for a patent and in due course one was granted for his "one-bath" chrome tannage.

The tanned leather so produced was different in a number of respects from any commercial leathers then on the market. For one thing, out of the tanning bath it had a robin's egg blue color. Another remarkable fact was that the tan was so set in the fibre that it would not wash out—in fact, to test whether his leather was full-tanned Dennis used to boil it. To produce a leather unaffected by water was an achievement. His chrome-tanned leather was far more porous than vegetable-tanned leather, and therefore in shoe leather it would allow more ventilation to the feet. For both shoes and gloves it was a boon to have a leather which would dry out soft after becoming wet by water or perspiration. It could be oiled and dyed in suitable manner to resemble practically any finished leather made by the old methods of tanning. The time of tanning was cut down by weeks for medium weight leathers and in the case of sole and other heavy leathers by months.

Having reached this point, what could Martin Dennis do with this new method of tanning? He put it up to his employers as a process which would enable them to develop new leathers, far in advance of anything then obtainable, resulting in a further expansion of their business. His uncles proved to be conservatives who could not see the profit in pioneering in such a well-rooted industry. No argument or persuasion could arouse their enthusiasm.

With faith in his new discovery, Dennis resolved to go on alone and resigned his position. He moved to Brooklyn and rented a building in which to manufacture chrome-tanned, glazed kid shoe leather. He operated for almost a year, but lack of working capital forced him to close down. However, the year's production showed the leather to be all and more than he had predicted.

So deeply did he believe that chrome leather was to make its place in the tanning industry that he decided to attempt the manufacture and sale of a chrome-tanning compound. He turned at this time to his friend of school and college days, Dr. Harry E. Richards, who had helped him secure his patent. Together they formed The Martin Dennis Chrome Tannage Company, later shortened to The Martin Dennis Company. This was in 1893. In a small factory, down in the Ironbound District of Newark, they began production of a "one-bath" chrome tanning solution under the name of "Tanolin."

Dennis was in charge of manufacturing and selling. This really meant demonstrating the value of his process of chrome tanning. The established tanning trade was not eager to make any change with the result the little company could make but little headway.

These were dark and discouraging times. Dennis would cover the trade as best he could but sales were next to nil. He would be given plenty of opportunities to demonstrate, but always in the end he would be politely told that there was no demand for that kind of leather. Despite his supreme faith, Dennis would have quit in the face of such constant opposition had it not been for the equally great faith of Richards and his patient encouragement to persevere.

Dennis went abroad in 1896 and in 1898 in attempts to interest English and Continental tanners in chrome leather. He found even less interest abroad than in the United States. After seven years of continual demonstrating some headway was made with a few tanners who began to make grain finished chrome-tanned calf leather. It possessed a quality of remaining soft and pliable even should the shoe made of it become wet. This was in marked contrast to the popular French waxed calf leather. The public immediately recognized this merit and a demand for the leather began to grow steadily. The next step was in the development of cow hide grain leathers in all of the popular finishes and also patent leathers. Then followed skins for glove leather and shoe lining and finally much chrome-tanned sole and mechanical leather came into use.

Another notable scientific contribution made by Martin Dennis to the tanning industry was the development of a chemical puer for hides and skins to replace the disgusting and unreliable natural manure bates used from ancient times. This bating process, always one of the least understood of all the tanning operations, was, however, one of extreme importance to the quality of the finished leather. Painstaking chemical research had revealed the probability that its results were obtained through the action of enzymes, and it was Martin Dennis who first in this country prepared them in a stable, usable form for tanners' use. The filth and stench that accompanied the use of the old bates made easier the introduction of these materials, and today it is safe to say that only in the most primitive tanneries do the old dog manures find any use.

The third important development for the tanner in which Dennis took great interest was the problem of applying the most suitable oil in the most useful way to lubricate the tanned leather. The objective he sought here was to secure the greatest strength compatible with the desired finish. His work on this problem resulted in the use of oils in emulsified form to replace in a large measure the hand and drum stuffing of leather with straight oils and greases.

Martin Dennis was a militant missionary in the field of chrome tanning. The charm of his personality and his great sincerity over-balanced the vigor of his arguments and won for him a host of acquaintances, many of whom were true friends. Some who at first felt that chrome leather would never become a product of the industry, later turned their entire production into it. From very small beginnings the business founded by

Mr. Dennis and Dr. Richards has grown through the years until its products are to be found in use by tanners in all parts of the world. To their persevering efforts, for Dr. Richards always stood squarely behind his partner and to him due credit must be given for the ultimate success, the tanning industry owes in no small measure the introduction of chemical materials and chemical methods. Martin Dennis was a conscientious student of exceptional mental honesty. Once he was sincerely certain of all his data, he had the courage to back them and the tenacity to follow straight through. The task he essayed was a difficult one. Had he been less bold, less persistent, less sincere, he could never have accomplished his great pioneering work in the field of the oldest chemical process industry.

Harry E. Richards

Jacob Hasslacher

JACOB HASSLACHER

1852-1921

THE morning's mail, a great stack of letters and orders nearly two feet high, had been read, every piece of it, by five men seated around an oval oak table and shuffled by departments into half a dozen neat piles. With a quick, characteristic gesture, Jacob Hasslacher slipped his gold-rimmed pincenez from his nose to the thumb of his left hand, abruptly shoved back his chair, and glanced at the big wall clock. It was twelve minutes before nine.

"Is there anything else—special?" he asked.

This was the regular morning ritual at the headquarters of The Roessler & Hasslacher Chemical Company. The officers and their chief read all orders and all the correspondence so that each might be fully informed on all details of the business, a task that also included a visé of the replies to important letters. Afterwards there was a few moments' terse discussion of any important, immediate development. By nine o'clock, each one was back at his own desk and ready for the day's work.

On this particular morning—it was a snowy day between Christmas and New Year's in December, 1916—the sales manager said, "They have advanced the price of sodium cyanide to two dollars."

"They?" Mr. Hasslacher shot back. "What 'they' do you mean?" and when the sales head did not answer, he added, "You mean the same old speculators and second hands?"

"Yes," Philip Schleussner assented, "the same second hands."

"And 'advanced'?" continued Mr. Hasslacher, "what is the meaning of 'they advanced'?"

Scorn rang in his voice. He was a manufacturer and merchant, and he had no patience with speculation in chemicals; but Mr. Schleussner continued, "I mean, Mr. Hasslacher, that second hands are now asking two dollars a pound for sodium cyanide, and" he added, "getting it."

Roessler & Hasslacher, real producers, were still selling at 25¢, and more than that, they had signed up contracts with regular customers for the coming year at that modest price. Throughout the year 1916, the open market price had advanced steadily from 28¢ in January to 40¢ in March, to 50¢ in September, to 68¢ in October, to 87¢ in November, and during the closing month it had sky-rocketed to an end-of-the-year figure of $2. During this same time, the R & H price had been advanced from 22¢ to 25¢.

This happened in the thick of the wild war scramble for chemical supplies. The European conflict had not only blocked the usual channels of chemical trade; but it had also created an enormous, insatiable demand. Importers and dealers scoured the world for chemicals. Manufacturers labored twenty-four hours a day with makeshift apparatus and impromptu processes to increase their output and to produce new materials. And day after day, warring nations and great industries bid and bid and bid against each other for sorely needed chemical supplies. Overnight prices jumped upwards and up again the next day. Fortunes were made by second-hand dealers who speculated in chemicals they had never seen and whose names they hardly knew how to pronounce.

Throughout this chemical orgy, Roessler & Hasslacher had maintained the policy of supplying the chemicals they made to their regular customers at fair prices. This was true not only of cyanides but of formaldehyde, chloroform, hydrogen peroxide, and the whole list of their own manufactured products. More than this, they were still willing to write firm contracts with consumers who were old customers for the coming year at prices far below the mounting market.

Such discrepancies between their prices and the open market quotations were costing the firm what appeared, particularly to the sales forces, to be perfectly legitimate profits. They subscribed to Mr. Hasslacher's principle of "taking good care of their regular customers"; but, as Mr. Schleussner was at pains to point out, it was one thing to protect their friends and quite a different thing when the friends took advantage of them.

"The spread has become so big," he said, "that some of our contract customers are re-selling our materials at the market price. We have proof positive of this, Mr. Hasslacher, and you cannot blame them. The temptation is simply too great."

"If you are positive that they are re-selling, we shall sell them no longer. We are selling them 'for their own requirements.' That is a part of the contract, and they have broken that contract."

"If we do that," put in William Hamann, "we shall win more ill-will than all the good-will we have earned by holding prices down."

"I understand that perfectly," said Jacob Hasslacher. "That is why we must be absolutely certain that they are re-selling, and in no case should we sell more than the maximum set forth in our contracts. Let us all understand each other clearly. I have been in this business a long time, and I have learned two things well. Easy profits are hard profits. If you take care of the customer, he will take care of you. We are not losing money: we are doing the biggest business we have ever done in our history. But this war is not going to last forever; and when all these fly-by-nights are forgotten, we shall be making money still selling those same customers. So let's talk about this no more."

That was Mr. Hasslacher's way of doing business. No expediency could swerve him from an obligation. He recognized no excuse in fulfilling the exact terms of an agreement.

One year, he contracted with an American producer

for his supplies of a certain chemical, the foreign price of which broke so sharply that his supplier shut down his own plant, bought abroad, and sold to R & H at a handsome profit. When he learned of this, Mr. Hasslacher laughed and congratulated the seller upon his good sense. But he continued to take in every pound for which he had contracted at the price he had agreed to pay. Several years later, the circumstances were reversed; and when his fellow manufacturer came and asked him for some price concession on his contract, Mr. Hasslacher refreshed his memory, and that contract, too, was carried through to the last pound and penny.

The very essence of a chemical sales contract was to Jacob Hasslacher the certainty of supply at a known cost which it gave the buyer, and the assurance of sale at a known profit which it gave the seller. To tamper with its terms was simply to defeat its purpose, and he believed sincerely that many of our economic maladjustments arose from the insecurity injected into business by careless purchasing promoted by lax contracts.

His convictions were not less strong on the subject of business sincerity. The sales agent for one of the big alkali companies, at a time before these producers sold direct to contract buyers, once secured from him a year's contract upon the direct representation that the price for the quantity bought was the lowest offered. Mr. Hasslacher subsequently learned that another buyer was paying less for the same tonnage of caustic soda. He summoned the salesman to his office; demanded an immediate refund on all deliveries made; cancelled the contract forthwith; and issued instructions that in the future R & H bought nothing from that distributor.

Although in such incidents Jacob Hasslacher appears stiff-backed in comparison with the give-and-take policies of the chemical trade since the World War, nevertheless he was never arbitrary in his position because he always stood firmly upon principles that all who did business with him soon learned and could clearly understand. In fact, he abhorred misunderstandings and insisted that all matters of importance be set forth in black and white. Once that was done, he refused to recognize any blends of grey. He believed thoroughly in taking the best possible care of customers. As an example, the price of cyanide rose to $3 during the war; but no old R & H customer ever paid more than 35¢ for his legitimate supplies. He did not, however, subscribe to the facile doctrine that the customer is always right.

Yet above all else, Jacob Hasslacher was a quick thinking merchant. It is in those words that he is characterized by a man who, under his tutelage, grew from office boy to director in the R & H company; and all who associated with him agree that his most notable talent was his accurate commercial insight into all chemical affairs.

He backed this talent with exceptional concentration of purpose; business courage of a high order; and boundless physical energy. Throughout his active career, he repeatedly declined flattering invitations to serve as a director of banks, insurance companies, and outside industrial enterprises, saying always that his own business demanded his full time and his undivided attention. He was invariably receptive to all suggestions from his research staff, and he was always ready to risk money on either a new product or a new process. It is significant that when he died his safe deposit box contained besides R & H stock, little beside gilt-edge bonds. Until he retired, he was at his desk every morning at eight o'clock, and it is still remembered that long before subways and motors, he once went back and forth between the New York office and the plant at Perth Amboy, N. J., three times in one day, carrying a stenographer with him and dictating a great sheaf of correspondence and composing two important reports. Finally, he combined business thrift with personal liberality. Charcoal bags were in his day a "returnable container" for which there was a credit of $5 a hundred. Unfortunately, they were worth so little, so hard to keep track of, and above all so useful as work-aprons in the plant, that small wonder they were not as scrupulously inventoried as carboys and steel drums. A telegram from the charcoal supplier, begging for the prompt return of any and all bags on hand brought to Mr. Hasslacher's personal attention the fact that they were short about three hundred of these returnable bags. For that $15 he turned the office, the shipping room, the warehouse, and the Perth Amboy plant all upside down, yet that selfsame year, because the cost of living was rising fast, he took half of his personal share of the profits and gave it as a bonus to all the salaried employees.

Such a man was Hasslacher of R & H—a keen merchandiser of chemicals, scrupulously honest, dominating rather than domineering; a jolly, genial Epicurean who relished a good joke, a fine dinner, a vintage wine. He came to the United States in 1884 and the following year went into business with Franz Roessler. He died in 1921, and in 1930 the company they founded was bought, two shares for one, by the du Ponts.

His father, Georg Hasslacher, was a public official in charge of the spa at Bad Ems in Germany representing the Duke of Nassau. There Jacob Hasslacher was born in 1852. He went at first to the public schools in Ems and later to the Hadamar Gymnasium from which he was graduated in 1872. With some training in the sciences, he became a salesman, entering the employ of the Deutsche Geld-und-Silber Scheideanstalt vormals Roessler of Frankfurt-am-Main. Here the young salesman was given a thorough-going course of training in all branches of the business.

A few years before Jacob Hasslacher joined the forces of the Scheideanstalt—in 1871 to be exact—that company had been founded upon the interests and the staff of the important Roessler family which for generations had served the old German Free Cities as Masters of the Mint. Among the products manufactured was a fine liquid gold largely used by the ceramics industry in decorating chinaware. This item began to develop a promising volume of export to the

Director Alexander Schneider of Scheideanstalt, and Mr. Hasslacher, from a photograph taken in the "Gay Nineties."

R. & H. dinner in honor of Hugo DuBois: At the head of table, Dr. H. R. Carveth, Dr. L. H. Baekeland, Mr. Hasslacher, Hugo DuBois, Mr. Roessler, John K. Creevey, Wm. A. Hamann.

United States, and in 1882 young Franz Roessler was sent to America to start a factory and cultivate the American trade. The factory was started in Brooklyn; but the business did not come up to expectations. Recognizing that Roessler, an extremely capable technician, needed an able assistant in the sales work, the company selected Hasslacher to strengthen their American venture. In 1885, a year after he came to this country, Hasslacher with Roessler organized the company which bore their names. The Anstalt was a third and controlling partner, and when the partnership became a corporation in 1889, the German company retained a majority of the voting stock. The active management, however, rested entirely with young Roessler for the plant and young Hasslacher for the business offices.

A new factory was built over at Perth Amboy to produce acetone from acetate of lime and chloroform from the acetone, also potassium cyanide from the prussiate, and quite an extensive line of pottery and ceramic colors. This operation began in October, 1885, and later there were added tin oxide, the metallic cyanides, and peroxide of hydrogen.

It became the established policy of the young firm to import chemicals from Germany until such time as a sufficient demand had been built up to justify a manufacturing operation in this country. Thus, in the late nineties, they organized the Perth Amboy Chemical Company to manufacture formaldehyde, in which enterprise R & H and the Hiag firm of Konstanz were partners. In the meantime, they had perfected a process for the production of the double cyanides by fusing sodium with the prussiate of potash. About this same time, Hamilton Young Castner had developed a practical electrolytic process for the manufacture of metallic sodium. Rights to this process were acquired to operate and the Niagara Electrochemical Co. was formed with the English Castner-Kellner Alkali Company, the German Anstalt, and the American R & H equal partners. Roessler remained responsible for the plant operations and Hasslacher for the sales and business administration. Sodium also became the starting point in the manufacture of sodium peroxide, the base from which the hydrogen peroxide and the perborates were made. In 1925, the Niagara Electrochemical enterprise was consolidated within The Roessler and Hasslacher Chemical Company.

The first sales offices of the company were at 56 Pine Street, New York, but very shortly they moved to 73 Pine, and after twelve years to 100 William Street, which, with the Drug and Chemical Club on the top floor, was for many years, before the post war exodus of the chemical companies into the midtown area, the heart of the chemical district. It was after the war, indeed after the retirement of both the original partners, in 1921, that the headquarters were moved uptown to 709-717 Sixth Avenue, opposite Bryant Park.

From its founding in 1885 to the time of the World War, Jacob Hasslacher was active head of the company. While his immediate personal concern was the commercial end of the business, nevertheless he kept in close touch with plant operations, and all new projects and all important negotiations came to his desk. However, he had been bringing up carefully trained associates. While before the war the staff which chiefly assisted Mr. Hasslacher and Mr. Roessler was William A. Hamann and Hugo Du Bois in the commercial end of the business, John K. Creevey, attorney, and Dr. Hans Foersterling in the technical department, others came more prominently into the picture during the war period. Philip Schleussner became Sales Manager as other burdens fell on Mr. Hamann, and Dr. Hector R. Carveth succeeded Dr. Foersterling. Constructive leadership in research and production was furnished at the Perth Amboy unit by B. S. Lacy, Sterling Temple, and L. M. White and at Niagara Falls by E. A. Rykenboer, Earl A. Harding, H. N. Gilbert, and P. J. Carlisle.

The World War inevitably brought considerable expansion of manufacturing capacity to the R & H enterprise. The dislocation of world chemical trade also wrought an important and far-reaching change in the set-up of the corporations. Throughout a busy quarter-century, the policy of manufacturing chemical specialties at first imported, as soon as American demand justified the capital investment in plant, had gradually, but definitely, brought to The Roessler & Hasslacher Chemical Company a great measure of independence from their foreign connections. While relations with their foster company, the Deutsche Geld-und-Silber

Scheideanstalt were always upon the most friendly basis of mutual cooperation, nevertheless both Jacob Hasslacher and Franz Roessler cherished this independence. They not only adopted the policy deliberately formulated to promote it, but also on more than one occasion they had protected it by insisting that the American business, both in plant and office, be placed unreservedly under their management. The wisdom of this policy was rather spectacularly proved when the war abruptly cut off European chemical supplies. R & H was able to carry on, even to advance, under conditions that seriously handicapped many old firms wholly dependent upon imported chemicals.

Their independence thus vindicated, the heads of the company, a year before the United States entered the conflict, felt justified in seizing the opportunity to make that independence legal and financial, as well as chemical and managerial. Accordingly, they sent a representative to Germany, commissioned to purchase at a fair price any stock in the American company that members of the Anstalt group were willing to sell. They were able to buy sufficient shares to secure control.

After the United States entered the war, when the validity of this stock purchase was being questioned, young Joseph Choate, as attorney for the Alien Property Custodian, visited Mr. Hasslacher at his summer home. They talked for a long time on the broad verandah with its superb view of the rounded shoulders of the Catskill Mountains that Jacob Hasslacher loved so well. At the time he was recuperating from an attack of the same illness which four years before had forced him to curtail his business activities, and he had just received word that his oldest son had been transferred from the U. S. Engineers to the Chemical Warfare Service. He mentioned neither fact, for characteristically he disdained to mix sentiment with business. Mr. Choate was endeavoring to establish the motives of that stock purchase a year previous.

"Naturally," said Mr. Hasslacher summing up, "it was a friendly transaction. They had furnished the original working capital that enabled us as young men to get in business for ourselves. More than that they had trained us in the business. For years they had given us good advice and technical assistance. For years we had bought from them and sold for them. Among them we had several close, personal friends. But, Mr. Choate, there was nothing friendly about the negotiations as to price and terms. That was all strictly business, though I do not mind telling you now that we would have paid more for the stock, if we had to, and we would have liked to have bought more of the stock, if they would have sold it."

Not till two years after Mr. Hasslacher's death did the Alien Property Custodian concede the legality of that purchase. In 1918, the shares owned by the Anstalt interests were taken over by the Alien Property Custodian. These shares, with the consent of the Scheideanstalt, were sold several years later to a group of R & H officials who in turn distributed them to employees and the other American stockholders. The Company thereafter continued to be one hundred per cent. American owned.

On January 1, 1920, Jacob Hasslacher laid down the reins. He was far from well, and after so many years of strenuous service, he sought rest and the restoration of his old super-abundant health. In earlier years he had never been sick, and it was a long-standing family joke that he recognized but two illnesses, pneumonia and broken bones. Yet he never fully recovered, and fourteen months later, on March 15, 1921, he passed peacefully away at his home in New York City.

Throughout his active life, business controlled but did not monopolize his interests. So warm-hearted and generous a man could not but devote a great deal of time to his friends and family, and also during the later years of his life particularly, he was a moving spirit in the organization and administration of Elka Park, a residential reserve in the Catskill Mountains, where for many years the Hasslachers had their summer home. Here and at his residence in New York, he delighted hospitably to gather together his personal friends and business associates at dinner parties and all sorts of informal gatherings.

In all his hospitalities, Mrs. Hasslacher, who before her marriage had been Miss Elizabeth Fleck, most charmingly played the hostess. They had a large family: two sons, George and Carl, and four daughters, Agnes, Antonia, Thea, and Emily. Always they were a closely-knit family group. He was a devoted father, playing and working with his children, going off with them on long walks, visiting museums, taking them to the theatre and opera, entering into their interests, training them, encouraging them.

Christmas was the great holiday in the Hasslacher family, and every year at this season all of the stray bachelors of their acquaintance were gathered together into the family festivities. During the very last Christmas season that he lived to enjoy so thoroughly an incident happened which made a deep impression on the men of the R & H organization who took part in it. At that time of the year, the sales staff had been called together from all over the country for a conference at headquarters. Learning of this, Mr. Hasslacher sent word that he wanted "his boys" to come up and see him.

Headed by Philip Schleussner, the managers of the various R & H sales offices from all over the country went up to his home. There was Alfred Stepan from Chicago, August Goerner from Boston, Carl Dittmar from Cleveland, Milton Kutz from Philadelphia, and William Stoll from San Francisco. Sitting in an invalid chair, he received them by an open fireplace.

He plied them with questions about customers that he knew well and conditions with which he had once been so closely in touch. He gave good advice on their problems and laughed heartily at their jokes. Plainly, he was itching to be back in the thick of the sales battle. It was quite plain to all of them that only a miracle could again make him their active commander-in-chief. Port and Christmas cookies were served and he proposed a toast for the coming year to a better sales record for R & H, and "his boys," gray-haired veterans most of them, drank a hearty toast to his own good health.

Amid cheery Christmas greetings, the little party broke up. Most of them never saw him again. All carried away an indelible picture of an alert, courageous leader; a warm, honest friend.

A gathering of the Roessler and Hasslacher children at the Hasslacher summer home at Elka Park, N. Y. Left to right, Ria, Hans, Fritz, Lily, Ann, and Emma Roessler; Emily, Thea, Antonia, Carl, George, and Agnes Hasslacher.

JOHN F. QUEENY

1859-1933

AT the insistent instigation of an enthusiastic young chemist in the Department of Agriculture a new kind of law had been introduced into Congress. It made the revolutionary proposal that definite standards, with tests for purity, be set up for all sorts of medicinal chemicals and drugs, and it provided that sellers of adulterated or sub-standard materials were to be punished by fine or imprisonment.

Even the most responsible people in the drug trade, men and firms of highest probity, sincerely believed that this proposed Pure Food and Drugs Act was a dangerous and unwarranted extension of governmental powers. They indicated politely, but pointedly, that Harvey Wiley, its sponsor, was a crack-brained visionary. Important technical and professional societies solemnly resolved that such legislation was an invasion of the rights of citizens. Powerful trade associations worked against the law in Washington. The conservative and influential National Association of Wholesale Druggists, after a short, hot debate, referred the matter to its standing committee on trade practices. The diplomatic chairman of that committee, in his report at the meeting the following year, glossed over this inflammable subject of the Pure Food Bill. However, one member of his committee, the director of purchases of the Meyer Brothers Drug Company of St. Louis, felt strongly for this revolutionary law, and he had the courage of his convictions. John F. Queeny read a minority report.

With a slashing stroke of wit he broke the thin ice, which the chairman had so skillfully avoided, and plunged in. Very quietly, very briefly, in telling phrases he pointed out the grave public responsibilities of every purveyor of medicines. "When it is a question of life or death," he said, "there can then be no question of quality or profit." He named names and quoted prices to show the many subtle varieties of sophistication that were common in those days. His sarcasm seared the opponents of a law designed to prevent such abuses. "This law is proposed for the protection of the public," he concluded, "but believe me, gentlemen, it will be a God-send to every honest maker, every honest seller of medicinal products."

So vigorous a report, delivered so fervently, by a man known to everyone, a man whose personal integrity was never questioned, a man who represented the largest wholesale drug house in America, had profound and immediate effects. It marked the first turning of the tide of sentiment in the chemical and drug field towards the Pure Food and Drugs Act, and it was this support within the trade that eventually carried the measure through both House and Senate.

Yet when John F. Queeny stood up to make that famous minority report of his, he faced an hostile audience. For an employee, even as trusted a man as he, in as responsible a position as he held, this was a courageous thing to do. He was thinking far ahead of his time, for what his contemporaries could at first see only as a restrictive law, he conceived truly to be a constructive measure to benefit the entire business. His earnest sincerity, supported ably by his most winning personality, enabled his convictions to triumph over the misgivings and the opposition of his peers. It was an experience that was repeated time after time in the career of this pioneer maker of coal-tar chemicals in America.

Of all that sturdy band of rugged individualists who during the past century founded our great American chemical companies, none was so incorrigible an independent, so indomitable a fighter as John Francis Queeny, founder of the Monsanto Chemical Company. With his background of experience and acquaintance in the wholesale drug trade, it was not unnatural, when he determined to become a chemical manufacturer, that he should select the fine and medicinal products of coal-tar origin. He must have been aware of the technical difficulties involved in the manufacture of this complex group. He knew perfectly well that this was the chemical domain which the powerful German companies had made their own and which they were determined to dominate. As a buyer he was quite familiar with the ruthless methods they had long successfully employed to prevent the establishment in this country of any domestic production of coal-tar dyes, coal-tar aromatics, and coal-tar medicinals. He had first-hand information on the comparatively limited market for the chemicals he proposed to produce; and he was amply forewarned of the negative, but stubborn sales resistance which any newcomer would meet in this comparatively new medicinal field due to exacting demands for purity and the extreme conservatism of buyers. It would almost seem that John Queeny had deliberately picked out the most difficult of all chemical fields and the one in which success, even under the most favorable circumstances, would be longest delayed. Courage was one of his most outstanding characteristics.

He needed great courage, and all of his great perseverance as well, for he met every obstacle which his experience and foresight had anticipated with some altogether unexpected set-backs. Literally, he battled his way ahead. For when most thickly beset with difficulties and discouragements, his plan of campaign was always founded upon an aggressive offensive. He refused to compromise, even when he might so have won a snugly entrenched position, if compromise meant any sacrifice of his independence. First and last, he was an industrial warrior who never recognized defeat.

During his lifetime, among those who did not know him well enough to penetrate beneath external superficialities, he had the reputation of being impulsive and swayed too greatly by his personal likes and dislikes. Apparently this was so, because his decisions were made with the speed and finality of a thunderbolt, while his personal friendships were as firm as his commercial rivalries were uncompromising. Furthermore, he was an exceedingly human individualist with a hearty personality; open-hearted and open-handed; outspoken even to the point of brutal frankness. He hid nothing, for he had nothing to hide since neither deceit nor trickery entered into his make-up.

Nevertheless John F. Queeny worked for long years upon the well thought-out pattern of building the Monsanto Chemical Company upon the foundation of coal-tar intermediates produced by themselves from American crudes. Today his oldest associates in that enterprise cannot recall a single important chemical which once made was subsequently discontinued. In several instances they made materials before the market here was ready for them; but invariably John Queeny's foresight was in time justified. At the close of the World War, when the Monsanto interest was bought into the coal-tar plant in Wales, the objective was to acquire a British contact with world markets. This he clearly observed was closer and more practical than any American chemical house might expect to make in many years, and he anticipated that such a contact would be a powerful weapon to keep the German manufacturers from regaining a stranglehold upon international chemical trade. No capricious executive reasons in this fashion nor leaves such a record of persistent accomplishment. In fact, the sinewy virtue of persistence was one of his most distinguishing attributes.

As a young boy John Queeny was suddenly confronted with a sink-or-swim crisis. If that little lad of twelve had not been born with self-reliance and tenacity, no other gifts of nature would have brought him safe to harbor in the port of success. His father was a wealthy, retired contractor, and John Francis was the eldest of five little Queenys. He had just finished grammar school and little thought had been given to his future for apparently he had before him eight or ten years of school and college. This pleasant, indefinite program was turned to ashes in the Chicago fire. In that conflagration his father's comfortable fortune vanished. Young John Queeny must go to work.

Chicago business was thoroughly disorganized. Good jobs were not easy to find. It was not, therefore, the romance of pharmacy, nor the lure of chemistry, nor even a burning ambition to become a great chemical industrialist, which started John Queeny upon his career at eight o'clock on the morning of March 28, 1872, in the wholesale drug house of Tolman & King. It was simply an accident, for as he himself delighted to relate: "I took my first job as an office boy in a drug concern at $2.50 a week, just because it was the first one I happened to run across."

From office boy young Queeny graduated into a

"runner for shorts," an arduous occupation in the pursuit of which he drove a light express wagon about the Chicago Loop, going from one wholesale druggist to another, collecting a twelfth of a dozen of Jones Cough Mixture; a half dozen bundles of licorice root, or whatever odds and ends his house needed to fill that day's orders.

It is a tradition of the Chicago drug trade that young John Queeny could make the rounds over to Kinzie Street, on to West Washington, and back to Wabash Avenue in a record time that has been lowered only in these days of motorcycle delivery. Years later when this early honor was unexpectedly thrust upon him at a banquet, he confessed with glee, "Oh, that was easy. I knew all the traffic cops, and they gave me the breaks at the corners. Besides I wore out more tires—iron ones, not rubber—than any other boy on the job."

Tolman & King were succeeded by John A. King & Company, whom Morris & Plummer later bought out. Young Queeny survived these changes in ownership, and in eleven years had worked up from office boy at $2.50 a week to a desk in the city sales department at $18. Then he was offered the position of buyer for I. L. Lyons & Company in New Orleans. Here he spent the next ten years, going thence to St. Louis in a similar position with Meyer Brothers Drug Company. Save for a three-year interlude when he was in New York with Merck & Company, he remained with Meyer

Brothers until he resigned to devote himself wholly to his own business.

In 1899, three years after his final return to St. Louis, Mr. Queeny fared forth on his first chemical manufacturing venture. It was an exceedingly inauspicious beginning. Had it been successful, it is not unlikely that today the Queeny name would be associated with heavy chemicals and the birth of the American manufacture of synthetic organic chemicals might conceivably have been postponed, as it was in the case of dyes, till the period of the World War. For John Queeny's first chemical enterprise was the refining of sulfur.

From his position as buyer for Meyer Brothers, he logically deducted that since St. Louis was the heart of the industrial Middle West, low-cost freight from the then newly developed sulfur mines of Louisiana made this the focal point for the refining of this material. He outlined his plans to the head of the house of Meyer and received his consent to conduct, under a paid manager, a sulfur refinery in East St. Louis while still retaining his position as head of their important purchasing department.

Accordingly, the manager was engaged; a building across the river rented; machinery installed; and in due course, six cars of crude sulfur were shunted alongside. By this time Mr. Queeny found that he had paid out nearly all of his six thousand dollars of savings.

At last the eagerly awaited day when the new plant was to come into operation arrived. A direct wire had been installed at Mr. Queeny's desk and all afternoon he impatiently awaited word that production was actually under way. Five o'clock came and no call on the plant telephone. Half past five, still no message. At a few moments before six, the call bell rang; but it was the regular telephone not the plant wire. In breathless phrases the manager blurted out his message.

"Fire! . . . . . Sulfur ignited in the conveyor system. . . . . . Nothing left but the concrete foundations."

Mr. Queeny thanked him and slowly hung up the receiver. Hardly had it slipped onto its hook when again the bell rang. It was Mrs. Queeny.

"John, dear," she said, "have you forgotten that we have guests coming for dinner at seven?"

Man-like, of course, he had forgotten. But he remembered in a flash that this was to be a party that his wife had been keenly anticipating and he instantly rose to a very un-man-like sacrifice.

"I'm just leaving the office. I was waiting for word from the plant."

"How is everything? Did they get going?"

"Everything's fine. They got going about four o'clock—going like a house afire!" and he chuckled at his grim little joke. That evening he outdid himself in the congenial rôle of the entertaining and hospitable husband of a gracious hostess. Not till the next morning did he break the news of the fire that had wiped out their savings and blasted his hopes.

Like that earlier fire which by chance had driven him into the drug business, this second one did not wither his courage; and two years later, in 1901, when his savings had accumulated to a very modest sixteen hundred dollars, John F. Queeny ventured again into the manufacture of chemicals. In two important respects, however, this second chemical-making endeavor differed from his initial effort. To facilitate raising outside working capital he formed a company, choosing for its corporate title his wife's family name of Monsanto. He set out, not to refine an old and important industrial raw material, but to produce one of the newest of the coal-tar group of fine chemicals.

The development of this group and its unbounded future possibilities fascinated John Queeny. He had witnessed the introduction of aspirin and phenacetin and noted their sensational successes from both the therapeutic and commercial points of view. His interest in the Pure Food and Drugs Act had prompted a special study of methyl salicylate, synthetic oil of wintergreen. When saccharin was introduced, almost coincident with the early development of the soft drink business in this country, he sensed a real opportunity, and he promptly went to his employers to point this out and to urge them to enter chemical manufacturing in this field. Such a proposal was contrary to every tradition of the house; but again permission was given to the chief buyer to engage in this business venture on his own time outside of office hours.

The saccharin market in this country was tightly in the grasp of the German manufacturers, so arrangements were made with the Swiss firm of Sandoz to supply the necessary intermediates and through their good offices a young Swiss chemist came over to America to take charge of the manufacturing operations. Thus Dr. Louis Veillon joined John Queeny at the very birth of the Monsanto Chemical Company. Theirs was a long, close, intimate association, broken only by death; and of each it is said that he trusted and relied upon the other as upon no other man. One of Mr. Queeny's strongest points as an executive was his unwavering confidence in certain key men of the technical staff. No chemist himself, he devoted his energies and intelligence wholeheartedly to sales and general administration, relying implicitly first upon Veillon, and later as the organization grew, upon Gaston Du Bois and Jules Bebie, H. O. McDonough and Lloyd Francis Nickell. Nothing could shake his faith in these men; and this well placed confidence, together with his infectious conviction of success, were undoubtedly important elements in the long up-hill struggle to establish coal-tar chemical manufacturing in the United States.

Assured of a process, of raw materials, and a competent production man, John Queeny set out to secure working capital and a plant. The start was made in what was little more than a glorified shed with an oil stove as the chief piece of equipment. Mr. Queeny furnished $3500 of the capital, his friend John Rossiter, $500, and Jacob Bauer of Chicago, president of the Liquid Carbonic Company, manufacturers of soda fountains and syrups, $2000. Two years later, in

1903, Thomas Wright bought a substantial minority interest in the young concern, investing $25,000, and his son Ralph Wright joined the staff in the sales office.

In the spring of that same year, Mr. Queeny made his first trip to Europe. He sought personal contact with the Sandoz people and was on the lookout for other products and processes that might be added to the manufacture of saccharin. Arrangements were made for a young Swiss chemist, Gaston Du Bois, to act as interpreter for him. They met on the docks at Rotterdam, and Mr. Queeny promptly turned over all his cash to his youthful guide with instructions to take him to the best hotel and to pay all his expenses. Swiss frugality triumphed over these blanket instructions, and they went to what Mr. Du Bois describes as "a very good, but not very expensive inn."

For three weeks they were together day and night, and John Queeny was so won by the character and ability of Gaston Du Bois that he persuaded him to come back to St. Louis with him "to visit the great St. Louis Fair and incidentally to perfect and install a process for making vanillin." The arrangement worked out perfectly, save only that Gaston Du Bois became so deeply immersed in the vanillin problem that he only visited the Fair for a few hours the evening of July Fourth. All the staff was working days, nights, and often Sundays. Mr. Queeny was carrying the double load of his work at Meyer Brothers and overtime in the Monsanto business, and at this time was drawing one hundred dollars a month for the responsibilities of finances which were serious and of sales which were difficult.

Production of vanillin, the most important of all the aromatic chemicals, proved to be a knotty problem, for the yields obtained in the laboratory fell off woefully in the plant. For three or four batches everything would proceed as scheduled. Then suddenly the yield would mysteriously drop. The money invested in research and apparatus was a serious item in the tiny budget of the young company. But Gaston Du Bois had won his confidence and Mr. Queeny's sole comment was: "Suppose you write out a full report, comparing each step of the laboratory and the plant operations. Then let's see what is to be done next." "What next?" came to be his invariable reaction to any check in their technical development.

In the meantime, saccharin was well in production; but ruthless competition from the Germans was raising a real crop of problems in John Queeny's own department of selling. The price had been slashed from $6 a pound to $3—eventually it was cut down to 60c—and more than this two other American makers, Heyden and Fries Brothers, had entered the lists under the German auspices. This domestic competition appeared in vanillin also, for Maywood, enjoying favorable German connections, soon began production. Monsanto, it was evident, would be choked to death, unless new products were added. Accordingly after vanillin came (roughly in this order) chloral hydrate, caffeine, glycerophosphates, phenolphthalein, coumarin, acetphenetidin, salicylic acid, acetylsalicylic acid, and the various salicylates.

Every new product meant a repetition of the same merciless battle for the trade. For example, before Monsanto began making chloral hydrate the price had been 85c. Finding his costs well below this figure, John Queeny decided to set the competitive pace by announcing a reduction to 60c. Thereupon the imported material was promptly reduced to 18c or below actual factory costs. Monsanto withdrew from the market temporarily and the price was promptly put back to 50c.

Diversification of products saved the day, but Mr. Queeny early recognized that so long as the company was dependent upon foreign sources of supply for intermediates no real stability could be won and it would continue to be vulnerable at a vital point. Accordingly, he brought over the third of the triumvirate of Swiss technical men who share with him in the triumph of successfully transplanting to the United States the seeds of our coal-tar chemical industry. Dr. Jules Bebie was engaged specifically to make toluene sulfonamid. He did so successfully. Thus, when the World War came a few years later, Monsanto was in the unique position among American saccharin producers of being able to manufacture in their own plant from the raw material to finished goods. With the War came also the shutting off of chemical supplies from Germany, upon which other makers in this country were accustomed to rely, and at the same time the cutting down of sugar supplies, so that the saccharin prices went up rapidly.

Like others among our chemical pioneers John F. Queeny believed in the letter of a contract. Therefore he fulfilled his obligations to the last delivery order, and while speculators were making fortunes, he endeavored to hold prices down for the benefit of his regular customers. It is an interesting commentary upon human nature that it was the tobacco people, who before the War were the first to support him as an American producer of both saccharin and vanillin, who were also, after the War, the most loyal of all his customers.

Naturally the period of the World War was a time of great activity and important expansion for Monsanto. The company was the only American-owned-and-operated producing unit in the field of coal-tar medicinals. Their experience was not only invaluable to their own development, but they were also an important aid to the Government in securing needed supplies of both finished medicines and raw materials for the making of dyes and explosives. One good example of the chemical services rendered is furnished by the manufacture of phthalic anhydride, which had been previously imported from Europe for the manufacture of phenolphthalein. On the outbreak of hostilities, Mr. Queeny immediately bought by cable all the available supplies that could be found around the world including 900 pounds found in Japan, some of it in one ounce tins. This but postponed the crisis in manufacture, and work was immediately begun on the old, rather unsatisfactory process of making phthalic anhydride by the oxidation of naphthalene in the presence of sulfuric acid and mercury. When the Government announced the perfection of the Gibbs process by the direct air oxidation, Monsanto was ready to translate the laboratory reaction into commercial production.

Mr. Queeny had the solid satisfaction of seeing the company he had founded emerge from the War in the position of being essentially a self-contained chemical manufactory. In 1917, the Commercial Acid Company plant in East St. Louis was purchased, making the company independent of outside supplies of sulfuric, muriatic, and nitric acids. The electrolytic production of caustic soda had been established in 1919, giving also adequate supplies of chlorine. Processes had been perfected in the laboratory and carried over to successful plant operation for the production from American coal-tar crudes of all the needed intermediates.

Monsanto growth did not stop when the stimulus of war demands was withdrawn. In 1920, with the deliberate intent of securing a place in world trade through the means of a close British connection, a half interest was purchased in the old chemical works of R. Graesser of Ruabon, North Wales. Established in 1867, this plant was the world's largest distiller of natural phenol, and though the original purpose of the alliance was seriously upset by the world-wide depression, nevertheless, Monsanto has since taken over complete control

The original Monsanto plant located at Second and Lafayette Streets, St. Louis—a photograph taken about 1903, showing the original traffic department—"Gus" and his mule-drawn delivery wagon.

Edgar M. Queeny, taken in his office, adjoining his father's, where he became President, 1927.

and is now manufacturing most items of its regular line in Britain. Since the War these have been augmented chiefly by the addition of a series of industrial and rubber chemicals.

Since the War too, expansion by consolidation has been carried forward in other directions. This program was initiated by John Queeny and before his death in 1933, the Merrimac Chemical Company, oldest and largest New England maker of standard industrial chemicals and the Rubber Service Laboratories, producers of chemical specialties for the vulcanizing and compounding trade, had both been joined to the Monsanto organization.

These consolidations and the firm establishment of Monsanto in their British branch occupied Mr. Queeny during the closing years of his life. He spent much time in England and Wales, for in 1927 his son Edgar had succeeded him as president and taken over the burdens of active management. Few men indeed have rounded out their own careers more completely and with greater inner satisfaction than John F. Queeny was able to do. He saw as an accomplished fact his long cherished ambition of a self-contained, widely diversified Monsanto Chemical Company. Secretly he must have loved a good fight and he had after years of hard commercial battle won a lasting victory. He placed the active control of that great company in the hands of his only son, and he lived to know that those younger hands were firm and capable.

Until within two months of his death on March

19th, 1933, Mr. Queeny was quite active in business. Thus he was almost the last survivor of the generation of great chemical individualists, great personal industrialists, executives of that distinctive and independent stamp which are no longer to be found in the close-meshed organizations of our great chemical corporations. His passing, as CHEMICAL INDUSTRIES pointed out at the time, "snapped a link with the colorful and illustrious youth of the American chemical industry."

FRANK SHERMAN WASHBURN

1860–1926

DURING one of those brief spells of very hot weather that are so often visited upon New York during September, three loyal alumni of Cornell spent the week-end at Rye laboring for their *Alma Mater*. In order to clear away the preliminaries, Frank Washburn, as chairman of the fund-raising campaign in the Metropolitan area, had carried off to his country home (for Rye was then well out in the country) the two best workers on his committee.

After an early breakfast, they went right to work. It was the routine task of dividing up territories, selecting captains, assigning good workmen and men of special influence or acquaintance to the different teams, allotting the prospects, estimating how the pledges would measure up to the quota—just the fussy kind of detail work that quite literally gives one a headache. Lunch was sent into the library. Dinner was rather hurried for there was still much to be done.

It was past midnight when the chairman-host shoved

back his chair and said; "Well, fellows, let's call it a day." Immediately, he jumped up.

"Come on," he exclaimed, "let's take a walk before turning in." He led the way, planning as he left the library that they return to the task at nine in the morning so as to finish in time for a swim before Sunday dinner.

They stepped out into the open. The night was still and clear, as soft and warm as a wool blanket; a respite, but no relief from the day's burning heat. They set off across the lawn at a brisk pace. After walking about a mile the chairman stopped short and turned around. "Now," he announced to his wilting committeemen, "let's run awhile."

After a driving day, "Now, let's run awhile"—that was Frank Washburn's way.

He was, as one of his long and close associates has said, "a man who was truly a glutton for work," and he indulged that inordinate appetite till it became almost a vice.

By profession Frank Washburn was a civil engineer, a capable, conscientious consultant with a flair for heavy construction work and a brilliant business imagination. By inclination he was an adventurous pioneer, a sort of twentieth century industrial Daniel Boone, always hunting, restless and alert along the outer fringes of the business frontier. By nature he was a reticent, rather introspective man who overcame a great natural shyness only in his middle years; but who in the heat of his enthusiasms became a driving master, a perfectionist demanding the utmost of himself and his men, impatient of delay, utterly intolerant of laziness and stupidity.

Through a strange series of haphazard coincidences he became interested in nitrogen and closely connected with important hydro-electric interests. From this curious combination the American Cyanamid Company naturally evolved. Looking backwards thirty years the clarity of his vision of the nitrogen future appears to have been almost a miracle of prophecy, and yet the sound logic of the business reasoning with which he planned to fulfil that future reveals a business leader of rare genius. Few, indeed, of our chemical pioneers have at once so boldly dreamed and so wisely done.

Frank Washburn was born in Centralia, Illinois, December 8, 1860. His father, Elmer Washburn, was an outspoken exponent of direct action, a shrewd, dictatorial individualist of straight New England ancestry, who had filled two important, contradictory posts in the Chicago stockyards. He was a past president of the National Livestock Bank and a retired head of the Stockyards Police Force. Such a dominating parent unintentionally suppressed a sensitive son, so that as the boy grew up, he naturally began to display an unwonted independence. Father and son attempted several partnerships together, but never succeeded in reaching a lasting agreement.

Frank Washburn was professionally trained at Cornell where he received the C. E. degree in 1883. He did a year's post-graduate work in economics, history, and politics and then went to work for the Chicago and Northwestern Railroad. He was quickly promoted to Engineer of Bridges, later to Engineer of Lines North of Chicago, and in 1889 was given the difficult assignment of reorganizing a subsidiary, the Chicago Belt Line Railway. Here he was brought into close contact with several important railway executives, and that same year the Illinois Central and the Lake Shore joined with his own road in sending him to England to study the methods and economics of British railway operation.

At this point the young engineer, but five years out of college, seemed plainly destined for a successful career in transportation. But upon his return he resigned and plunged into water works construction.

He went to work on Purdy's Dam, up to that time the biggest construction job in the New York water supply system, and afterwards he was placed in charge of building the two dams and reservoir at Carmel, N. Y. This work completed, he took over the construction of the Third Avenue Cable Railway in New York City.

Shortly before this last construction work was completed Mr. Washburn married Irene Russell of Augusta, Georgia, daughter of a prominent southern family whose sister was the wife of the President of Nashville, Chattanooga and St. Louis Railroad. While working on the cable line he met Colonel Grace, head of the great South American trading house. These two apparently quite unconnected events were the first links in the chain that led him eventually to air nitrogen.

Even as a very young man Frank Washburn inspired confidence and he was sent to Chile by the Grace interests to overhaul their nitrate operations. Though the field was new to his experience, nevertheless he was able very successfully to apply what he had learned building dams to the problems of large scale handling of the bulky caliche, while his cable railway lessons were more than useful in the transport of the refined saltpetre from the high, steep-sided plateau to the water's edge. He scored a conspicuous personal engineering triumph in Chile and was forthwith sent to Central America commissioned to review the surveys for the proposed Nicaraguan Canal and to report upon the economic soundness of the project. His findings—both technical and commercial—have been confirmed by History; but even at the time, his reports were so clearly stated and so logically buttressed with facts that he came back to the West Coast, after these experiences in the southern continent, with a well established reputation both as an engineer and a business analyst. He also brought back from South America a lively interest in nitrates, supported by first-hand information of the Chilean fields and of the far-flung, well-organized British and American interests which marketed this prime ingredient of both explosives and fertilizers.

His services were immediately sought as chief engineer of the Bay Cities Water Co., which was building a new water supply system for San Francisco and a group of suburban towns. Here he engaged a young engineer, K. F. Cooper, who was henceforth to be closely associated with him in a number of enterprises. At this time too, he first came into contact with H. M. Goodman, and once the water works were completed, he joined with him to develop a line of mining machinery. He plunged into this work with all his accustomed vigor. Within two years the problems of design and production, which fascinated him, having been solved, he lost interest when the problems of merchandising became paramount.

His devoted wife, with a feminine dislike for the roving commissions that carried her engineer-husband all over the map and with a perfectly natural homesickness for her native Southland, undertook a delicate piece of engineering herself with the result that the family moved to Nashville and here Mr. Washburn became an associate of her nephew, Whitford Cole. This was in 1900, just at the time when young Cole, having branched out from the family's original railway activities into coal and coke and blast furnaces, began to turn to the development of hydro-electric utilities. An informal alliance was made with Henry Parsons, an influential New Yorker with interests near Sheffield, Alabama, and Frank Washburn, the practical engineer of the group, was sent out to explore the potential power sites of the region. During this period he was largely responsible for the organization of the Birmingham, Montgomery and Gulf Power Company (incorporated December 8, 1900), the Little River Power Company,

Frank S Washburn

Mr. and Mrs. Washburn, with their daughter, Elizabeth, in the garden of their home, "Greywoods," Nashville, Tenn., in 1916.

and finally the Alabama Power Company of which he was the first president.

Thus from the opening of the century to the outbreak of the hostilities in Europe, Frank Washburn was an outstanding leader in the development of hydro-electric utilities in the South. He visited every likely power site in the southern Appalachians. He examined into all the developed and undeveloped natural resources—timber, coal, limestone, bauxite, clays, mica, and what not—locating them and appraising each source of supply for its economic and geographic values. He made a painstaking study of the existing and potential industries of the entire region from the Blue Ridge to the Gulf of Mexico. His associates followed his recommendations almost implicitly so that he was the prime mover in the building, and later in the operation, of these important power developments.

All his studies, backed by all his practical experiences, focused his attention upon the Muscle Shoals of the Tennessee River. Here, so he convinced himself, was the finest power site in the whole territory. This interest in Muscle Shoals was the final link that led him to air nitrogen.

At Muscle Shoals the Tennessee drops nearly one hundred and fifty feet within thirty miles. Since at certain points the river sprawls almost to two miles width, these swift, shallow rapids are a serious obstruction to navigation. As far back as 1835 Alabama built a series of small canals along the edges of the shoals, and in 1870 the Federal Government took over and so improved the Muscle Shoals Canal that it became practical for small boats. It was, however, still insignificant as a means of river navigation in any sense competitive with rail transportation.

Realizing at once that while no plan to develop hydro-electric power at this location would be sanctioned that did not maintain the navigability of the stream, nevertheless any such elaborate project would double the costs of providing a sufficient, year-round head of water for a large electrical output, Mr. Washburn worked out a program whereby the Government would construct a larger canal; the Government and private capital would build jointly the necessary dams; the private interests would erect the power plant and transmission lines.

This proposal was heartily approved by the State of Alabama which had everything to gain. The Army Engineers sanctioned the plan as providing, at smallest expense to the Government, a maximum improvement in navigation on the most important river in the Southeast. Any hydro-electric plant at Muscle Shoals, if the Tennessee were to be maintained as a navigable stream, would require an initial investment so heavy that Mr. Washburn estimated that the carrying charges at six per cent. would have represented eighty per cent. of the costs of the developed power. Accordingly, despite the favorable aspects of the river at this point, the rich natural resources of the region, and the potential power market in the Birmingham district, commercial development by private interests could only become feasible with some such cooperation with the Government in construction of canals and such dams as would preserve, even improve, the navigability of the stream.

Based upon this joint program the Muscle Shoals Hydro-Electric Power Company was organized and Representative Richardson of Alabama introduced a bill into Congress embodying this proposal. With the support of Alabama and the sanction of the Army Engineers, approval of this measure seemed certain, but the bill dragged through several sessions. Conservation of our natural resources, first urged by Theodore Roosevelt, was at this time a live political issue and President Taft had very plainly indicated that he wanted, not "a lot of mill dam legislation, but one national water power law." In the meantime, active water power development had been begun on a non-navigable stream, the Coosa River, where a $10,000,000 investment was made by the Alabama Power Company.

As long as the Muscle Shoals development was in prospect, it was plain that a big industrial consumer of a large block of primary power would be almost essential to the success of so large a project. Accordingly, Mr. Washburn began a search for such an initial customer. Since his Chilean days he had been interested in nitrates and he knew intimately the fertilizer situation in the United States. The proximity at once to the big fertilizer market of the South and to the Tennessee phosphate fields did not escape him. Accordingly, when he heard of the discovery by Birkeland and Eyde, of the arc process for nitrogen fixation, he promptly sailed for Europe to investigate it.

A power industry that produced fertilizer materials fitted perfectly into the logic of Mr. Washburn's plans. Therefore, after he had determined that, under American conditions, the arc process was not a sound commercial enterprise, he went from Norway to Germany to examine into the possibilities of another electrical method of fixing nitrogen that had been announced during his negotiations with Samuel Eyde. This was the Frank and Caro's cyanamid process.

Though consuming only a fourth as much electrical energy as the arc process, the production of cyanamid was essentially a power operation. Moreover, it required large quantities of coke and lime, both abundant in the Southern mountains. Mr. Washburn convinced himself that this process could be operated profitably in this country, so he purchased the American rights; and on July 22, 1907, the American Cyanamid Company was organized.

As there now seemed to be but little prospect of immediate development at Muscle Shoals, Mr. Washburn and his associates, W. R. Cole, his partner, William H. Lindsey, A. H. Robinson, a leading Nashville banker, and Charles H. Baker of New York, turned naturally to Niagara Falls. But carbide, aluminum, carborundum, and graphite were already big consumers taking up the base load of the American company. He therefore found it advantageous to negotiate with the Ontario Power Company whose new plant was just in operation on the Canadian side of the Falls. A twenty-five year contract for power at $10.50 per horse power year was finally signed. This rate was far cheaper than the lowest possible cost of their own power developed in the South.

Construction was immediately begun, and on December 4, 1909, the first carload shipment of three hundred bags, 52,240 pounds, left the plant. K. F. Cooper was works manager. S. W. Mays and P. F. Ronan were the plant foremen. The two former had been with Mr. Washburn on his own consulting engineering staff. Today Mr. Cooper is vice president of the American Cyanamid Company and Mr. Mays, the general purchasing agent. Mr. Ronan is now superintendent of the Company's plant at Azusa, California.

The Niagara Falls operation expanded with amazing rapidity. The initial capacity in 1910 was 5,000 tons a year. By 1912 this had been increased to 12,000 tons; in 1913 to 25,000 tons; by 1915 to 50,000 tons.

Then came the War.

Long before we entered the conflict the nitrogen problem caused serious concern in Washington. Both for explosives and fertilizers we were obviously, even painfully, dependent upon Chile. The first chemical lesson of the World War had been that all General Staffs—not excluding the German—had grossly underestimated the explosive demands of modern warfare. At the same time we had discovered that the fertilizer demand for nitrogen grew quickly and became imperative in response to wartime needs for more foodstuffs raised, of necessity, on curtailed acreage with less farm labor. As early as 1915 the National Academy of Sciences appointed a distinguished committee to study the nitrogen problem. This committee, after investigating our probable nitrogen requirements, in event of war, visited Europe to explore the arc and cyanamid processes and to learn what they might about the Haber ammonia synthesis, for it was no secret that this process had saved Germany from nitrogen bankruptcy. In his reports for 1915 and 1916, General Crozier, Chief of Ordnance, had called particular attention to the nitrogen situation; and the Army and the Navy and the Department of Agriculture and the Interior, had all studied the problem. In May, 1916, Congress passed the National Defense Act which among other provisions appropriated $20,000,000 for the building of a nitrate plant or plants, the location of which and the process to be employed and the disposal of the surplus power or nitrates all being placed wholly in the hands of the President. It was under this National Defense Act that nitrate plant Number One to operate a modified Haber process worked out by the General Chemical Company and Number Two a cyanamid plant to be built by the American Cyanamid Company were constructed at Muscle Shoals.

So much has been done at Muscle Shoals since October 29, 1918, when that cyanamid plant began operating its first trial run, and so much more has been said and written, that the importance of that war-time achievement has been completely obscured. It was, however, one of the greatest and most successful of all the great war triumphs of the American chemical industry. In Washington, November 6 to 8, 1917, Mr. Washburn, Dr. Walter S. Landis, the Cyanamid Company's chief technician, and K. F. Cooper, vice president, in conference with the Ordnance Department, drew up the rough specifications and raw material requirements for a 200,000 ton plant. Contracts were signed in December. Ground was broken in February. The foundation was laid in March. October 14, 1918, the liquid air plant was put in operation; October 21 the limekilns were fired; October 29, the carbide furnaces started.

Within a year, at a site out in the wilds, with no rail or road transportation to start with, amid extreme difficulties of getting both labor and materials, and under all the pressure of war conditions, the world's largest cyanamid plant was built with units for the production of both nitric acid and ammonium nitrate. Without these additional facilities, in a settled community with good transportation and during peace, the Company's best previous record had been a plant of one-seventh the cyanamid capacity put into operation in eleven months.

And the Muscle Shoals cyanamid plant operated successfully. The trial run performances of almost every unit of equipment surpassed rated capacities. The end product sought, ammonium nitrate, exceeded the government specifications in every respect save sulfate content which was .15 against .05 per cent. Considering the difficulties met and surmounted, it was an unparalleled feat of chemical plant design and construction.

Into that tremendous, patriotic effort, Frank Washburn threw all his great energies. He labored day and night, working literally fourteen to sixteen hours daily for weeks. He was justly proud of the results achieved, and the political hell-broth brewed after the Armistice at Muscle Shoals disgusted and saddened him. The unfounded accusations against his company and himself, the recriminations of conflicting interests, the political insincerities that developed, disillusioned him, but he never lost his faith in mankind. Nevertheless the injustice and ingratitude and duplicity of it all worried him and shortened his days.

In the meanwhile, the American Cyanamid Company—his company—had been firmly established. Already it had begun to diversify its products and so to widen its field of operations.

Back in 1910 they had started experimenting with the treatment of phosphate rock and the following year brought in Dr. Landis, then associate professor at Lehigh University, to investigate the production of phosphoric acid by new methods. Mr. Washburn sought shrewdly to add a line of phosphatic materials to the Company's nitrogen-bearing fertilizer. Both of these projects attracted James B. Duke whose tobacco interests had naturally brought fertilizers to his notice and who was already a substantial shareholder in the Virginia-Carolina Chemical Company. In 1914, when

At the Niagara Falls plant, 1908: *Major Berry, K. F. Cooper, Mr. Washburn, S. W. Mays, and G. A. Hendrie.*

Mr. and Mrs. Washburn, with their children, Frank S., Jr., and Elizabeth, vacationing at Highlands, N. C.

The original plant of American Cyanamid Co., at Niagara Falls, Ontario, Canada, as it was completed in 1908.

Mr. Washburn and Mr. Duke were both in Europe studying nitrogen fixation with the immediate object of bringing to America some cheap, practical method of producing nitric acid, they met in London, and Mr. Duke proposed an alliance of his phosphate rock and Mr. Washburn's cyanamid in the further development of Ammo-Phos. This new nitrogen-phosphorus fertilizer has been developed by Dr. Landis and successful pilot plant manufacture realized in 1912.

From this alliance grew the independent Ammophos Company, a plant at Warners, N. J., and a phosphate mine, acquired from the V-C subsidiary, the Amalgamated Phosphate Company, at Brewster, Florida. Before the new Warners plant was in production, the Cyanamid Company, by an exchange of stock with Mr. Duke and the Virginia-Carolina Chemical Company, took over Ammo-Phos completely, and in 1923 the V-C holdings which were sold to Mr. Washburn's old associates in Cyanamid.

Throughout his life, Frank Washburn followed the maxim of Theodore Roosevelt that it is better to wear out than to rust. In the end he broke down his overdriven body, but till his death the vital spark of his enthusiasm burned brightly. Had he lived another ten years it is likely that he would have explored some other new industrial field, for he was incorrigibly the pioneer. He was active in the development of twenty-four companies, with a majority of them since the very inception of the idea out of which these various enterprises grew. Eighteen of these companies became conspicuous successes. There were but two flat failures in his record.

Mr. Washburn once gave to his son a piece of advice which, based on his practical experience, summed up his philosophy of modern life. Having followed the father to Cornell and just back from active service in the Navy during the World War, the son was wrestling with the problems of a young man on the verge of his business career. He was attracted to the automobile industry, but his father said:

"It is too late. The automobile industry has grown up while you have been at school and college. Look at tomorrow, not today, when you pick your life work. Start with an industry in its infancy—airplanes or chemicals, if you will—it is only by growing up with a big industry that a young man in this country now can find a really big opportunity."

HERBERT H. DOW

1865-1930

DURING the winter term of 1887-88 one of the six seniors at the Case School of Applied Science went to work to prepare his thesis on "The Chemical Uses of Fuel in Boilers." Nobody—least of all, that serious, studious young man himself—guessed that he had set forth on a path that was to lead straight to the presidency of a vast corporation with 250 acres of great plants all built upon his skill as an engineer and his ability as an executive.

He was a quiet, but friendly youth. At a larger college his husky, rangy frame would have delighted the eye of some discriminating football coach. In the small, hard-working classes of Case of the early days, his alert, curious, retentive brain could not but impress his teachers. They recognized a promising student with independence and courage, honesty and determination, to support his exceptional mental powers. Within a few years these same teachers were to have the opportunity to substantiate their early recognition of Herbert H. Dow's capabilities and character. Several of them were among his earliest financial backers.

In preparing his thesis young Dow undertook a conscientious study of the chemical composition of fuels. While collecting a sample of natural gas near Cleveland, the driller called his attention to the exceptionally acrid taste of the brine from the same well. This raised questions outside the scope of his fuel inquiries, which most young chemists would have ignored; but his brisk curiosity was aroused and with characteristic energy he set out to find why that brine tasted so bitter. Analysis revealed an exceptionally high lithium and bromine content.

At this point another of Dr. Dow's characteristics early displayed its force. He knew that oil was selling at a dollar a barrel, and he quickly figured that there was about three dollars worth of lithium in a barrel of the brine. His commercial instinct, always keen and always stimulated by his chemical investigations, told him that that particular well was more valuable for its lithium than its petroleum; but his commercial instinct was sound and he recognized, at the same time, that the bromine content of this brine was even more important industrially than the higher priced, but less demanded, lithium.

Writing of these studies many years later he said: "I also did some development work on extracting the bromine from this sample of brine, and found that the bromine content was higher than in the brines of the wells in the Ohio River district, where bromine was being manufactured commercially."

Here lies the germ of the idea that grew into the

Dow Chemical Company. For these investigations, carried along while he was preparing his thesis on fuel, prompted a further study of lithium and bromine in natural brines, which took him to Pennsylvania, Ohio, West Virginia, and finally to Michigan. From samples collected on these trips he learned that lithium was present in greatest quantity in the Ohio brines, but quite absent from the Michigan brines; and most important of all, he discovered a maximum of bromine in brines from Canton, Ohio, and Midland, Michigan.

Continuing his account, Dr. Dow wrote: " In the fall of 1888 I accepted a position as professor of chemistry and toxicology in a medical college in Cleveland, where I had a laboratory and an assistant, and I utilized all my spare time in perfecting a process for extracting bromine from brine. This process offered better commercial possibilities than the only process I had been able to devise for lithium."

He is now fairly embarked on his great life work, and in the spring of 1889, having interested some local capitalists, he started a small bromine extraction plant at Canton, Ohio. Chiefly because these Canton capitalists were impatient, this venture was a commercial failure. However, the operation convinced Dr. Dow that the extraction of bromine by oxidation without evaporating the brine was chemically practical. He saw in the over-production of salt and the passing of the boom lumbering operations, with their low-cost fuel of scrap and sawdust, the opportunity for his process to replace economically the old evaporation plants that willy-nilly produced both salt and bromine. Accordingly, he returned to Cleveland, and after working out some refinements in his process, entered into partnership with J. H. Osborn of the National Carbon Company, Cleveland. They leased a brine well, situated square in the middle of Main Street, Midland; and in the fall of 1890 a tiny plant, with a crew of three men, one of whom was Herbert Dow himself, began operating twenty-four hours a day.

Midland was, at the time, the largest producer of bromine in the United States, and but a few months proved that unless the operation might be conducted on a larger, more economical scale, the little new plant could not compete with its established rivals. Accordingly, two more partners were taken into the Midland Chemical Company. Within the year it was determined again to expand. The partnership was turned into a corporation with $100,000 capital stock, and the effort was made to raise $27,500 in cash. Only $10,000 worth of stock was sold; but young Dow evidently lost financial control of the enterprise and seemed destined for the fate of so many inventors since at the organization meeting held in Midland on August 17, 1892, W. B. Remington was elected president; Thomas Percy and J. H. Osborn, vice presidents; W. W. Cooper, secretary, and B. E. Helman, treasurer. H. H. Dow was engaged as superintendent and the business affairs placed in the hands of the new treasurer, Mr. Helman.

But the young superintendent continued his chemical investigations. In the first step of his own process he replaced bleaching powder with chlorine as the oxidizing agent, and a year later he erected the first electrolytic chlorine plant in America right beside the fine new bromine plant. The current had been turned on but an hour when a tremendous explosion wrecked the new building and seriously damaged the bromine plant.

The directors, all of whom lived in Cleveland, were summoned post-haste to Midland to attend a special meeting, almost an indignation meeting. The inventive superintendent was instructed in no uncertain terms to produce bromine and to cease forthwith his silly and costly experiments; and the official records of the Midland Chemical Company record the verdict of no more expansions, a decision that was literally carried out, so that for several years all earnings were paid out in dividends.

Though even his minority stockholder's share of those dividends was sufficient to satisfy most ambitious young chemists, nevertheless dollars did not drive new chemical ideas out of Herbert Dow's head, and no power on earth could have kept him from testing out those ideas. As a student in college he had said: "I would rather make $3,000 a year on my own than earn $10,000 a year working under the direction of someone else." It was not a boast, but a principle; and so, in 1895, he is back in Cleveland again with a proposal to exploit further the chemical possibilities of brine by manufacturing electrolytic chlorine and its derivatives. This persistent pursuit of chlorine by electrolysis, even after the disastrous explosion, reveals another dominant trait in Dr. Dow's character. Once convinced of the soundness of a chemical idea, neither time, nor effort, nor cost counted with him as against ultimate success. Many years later he persistently ploughed through great snowdrifts of discouragements to the commercial production of magnesium metal and the commercial marketing of light alloys.

In forming the Dow Process Company to exploit electrolytic chlorine, Dr. Dow was most fortunate in his associates. Among them were his original partner, J. H. Osborn; his classmate, James T. Pardee; and two of his teachers at Case, Cady Staley, the president, and Albert W. Smith, a professor of chemistry. Out of this partnership grew the Dow Chemical Company. For after running an experimental plant at Navarre, near Canton, Ohio, about a year, this electrolytic operation was moved back to Midland. Shortly thereafter the Dow Chemical Company was incorporated.

Financial resources were strengthened by the sale of stock, and at the time of the election of the Board of Directors, May 18, 1897, there were fifty-seven shareholders of record. On that first board, in addition to Dr. Dow, his classmate, James T. Pardee, and his old teachers, President Staley and Professor Smith, were A. E. Convers, a manufacturer of tacks who upon Dr. Dow's death became Chairman of the Board; two bankers, Charles A. Post of Cleveland, and William L. Baker of Midland; two important manufacturers, S. T. Wellman (steel) and G. E. Collings (woolen goods). Most of these men were, for many years, closely identi-

Herbert H. Dow

Herbert H. Dow at the time his chemical projects were coming into his own and he was well established at Midland.

fied with Dr. Dow's struggles and triumphs, and at the time of his death, thirty-five years later, three of them were still active in the Company's affairs.

Three years later the Dow Chemical Company took over the plant and bromine processes of the Midland Chemical Company, and in 1902 a new subsidiary company was formed, under the old Midland name, to manufacture chloroform by a process originally worked out by A. W. Smith and William O. Quale. Until 1914, when it was purchased by the parent organization, this company remained a separate corporate entity.

With the introduction of the chloroform operation, Herbert Dow launched forth boldly upon the career that began with his student-day analysis of natural brines and ended with the commercial production of over two hundred and fifty chemical products drawn fundamentally from Midland brine. Beneath the story of the technical and industrial exploitation of brine lay the manufacturing and merchandising philosophy of this great chemical industrialist.

One beautiful spring afternoon, six months after the close of the World War, sitting in his beloved garden, Herbert Dow expounded this philosophy of his to me. It was a warm, sunny day and thousands of young apple trees, all pink and white with bloom, stretched away in long, flowing rows from the little hillock where we sat in comfortable rustic chairs. The morning had been spent inspecting the plant, then in the throes of vigorous readjustment from a war to a peace basis. Some operating departments were cold and silent;

others were undergoing obvious revamping; still others were orderly running at capacity. Dr. Dow showed everything and explained everything—why, for example, the chlorbenzol operation was "down" and wherefore he was planning a new phenol research. Plainly he had made the important decisions of policy and was already working out definitely his plans for the new era in American chemical manufacturing. His conclusions reached, his program drawn up, he was in that comfortable frame of mind that confidence in the future and in himself assured; and he was prompted by the peace and beauty of the scene before us to speak most frankly. During lunch we had been talking about chemical price quotations, and I had been enlisting his support of a movement openly to publish contract prices. As we sat there in the garden he returned to this subject.

"Price policies and programs of production seem to me the two fundamentals of chemical manufacturing about which we cannot do too much thinking. Both should be carefully planned. When they are allowed to become haphazard, they cause a great deal of trouble both for the individual company and the industry as a whole. Because we can make a certain chemical is no reason at all why we should do so. Because a competitor fixes a certain price, be it high or low, is no reason why we should meet it.

"We stand at the beginning of a great period of chemical expansion. If we, the chemical industry, are to make the most of this opportunity, it is necessary for us to make chemicals of higher and higher grades, more uniform and more dependable, at lower and lower prices. Only in this way can our products become the chemical tools for other industries. These are the plain facts of the situation. We must make our plans and policies square with those facts.

"Here at Midland our job is to make chemicals out of our brine. I want to see us extract every possible chemical opportunity out of that brine, and I will not be satisfied until we do not buy a single outside raw material, except," he added, with a smile that wrinkled up his friendly blue eyes, "say, a little C. P. sulfuric acid for laboratory work. But I am not at all interested in making anything, except to salvage a by-product or a waste, that we cannot make cheaper and better than anyone else is now making it. When we do make it better, I want to sell it as cheap as anyone. When we make it cheaper, I want to pass that advantage on to the consumer at once."

That philosophy, translated into action, is the story of the Dow Chemical Company. The development of the Dow Process for making phenol from chlorbenzol rather than by the sulfonation of benzol is a neat example of the principles and the practice. Dr. Dow early sensed that the plastics industry would be a growing market for phenol, that the cost of this item was an important factor in the price of finished plastics, and that lower prices would open up new fields for these then-new synthetic materials.

Having operated the sulfonation process during the War he also knew that the determining element in the cost of phenol at Midland was the price of sulfuric acid. The chlorbenzol operation was within the scope of his program to develop to the uttermost his own raw material resources. In spite of the knowledge that the Bakelite Corporation, the largest consumer, was building its own phenol plant, he went ahead with this development work, because he was confident that the costs of his new process would be lower. When the Dow phenol operation came into production in 1925 the price dropped from 36¢ to 21¢, and Bakelite, while holding their new plant in reserve, contracted with Dow for their requirements at a price substantially the same as their own making cost. Furthermore, in 1929, when the original phenol-formaldehyde patents expired and new plastic makers greatly increased the demand, Dow, instead of raising the price, deliberately lowered it to 11¢.

The chemical exploitation of brine, which began with ferric bromide (the first commercial product of the old Midland Chemical Company), proceeded quickly to potassium bromide, using the cheap potash liquors leached from waste wood ashes of neighboring sawmills. The electrolytic process brought with it both bromine and chlorine as well as caustic soda. Production of chloroform, a logical development to utilize the increasing output of chlorine from the fast-growing battery of electrolytic cells, brought with it both carbon bisulfide and sulfur chloride, which in turn led to carbon tetrachloride. This last was the first big tonnage Dow product. Its expanding market as a solvent led to constant increases in the capacity of the carbon bisulfide plant, so that when the rayon industry became a large consumer of bisulfide, Dow was in a strategic position to capitalize this new demand.

Chemical employment of the magnesium content of the Midland brine began with magnesium chloride, another early, big tonnage item that expanded enormously when the War shut off imports. Magnesium chloride led to magnesium sulfate (epsom salts) and magnesium carbonate. During the War development reached out into the organic field. Phenol and dinitrochlorobenzol were made for the Allies before our entrance into the conflict; and the use of bromine in the production of brom-indigoes was forecast when Dow became the first American manufacturer of synthetic indigo. Many additional brine wells were drilled and production of bromine and chlorine, essential for many munitions uses, was pushed. Mustard gas was made from ethylene and sulfur chloride. This was the first industrial production in the United States, and persistent tradition in Midland says that this Dow product was the only American-made mustard gas actually to reach the battlefront. With home-made phenol it was logical to make salicylic acid and the various salicylates, and the extension of the Dow line into coal-tar medicinals and perfume bases followed.

So this process of "extracting the chemical possibilities of brine" has continued, some of the most important developments being a new process for aniline, like the phenol process in that it is especially adapted to the

Dow operations: the production of the chlorides of magnesium and calcium by new processes; the synthesizing of acetic anhydride from sodium acetate and sulfur chloride. And finally, and especially interesting since they were Dr. Dow's pet projects during the closing years of his life, the extraction of metallic magnesium, which resulted in a series of extremely lightweight alloys.

This resumé of the practical working out of Herbert Dow's philosophy of chemical manufacturing, by omitting the details, creates a misleading impression. Just as the great modern plant, stretching a full mile along the Tittabawassee River, has blotted out the tiny wooden building in which bromine was first produced, so this catalog of triumphs obscures the hours of patient research, the days and nights of heartbreaking work in every new unit of that now gigantic plant, the bitterly fought sales battles. So too, the tall figure of a young chemical engineer with a black Van Dyke beard and bright blue eyes that snapped with boundless energy, is hidden behind the big, friendly, mature executive who was so often mistaken by strangers for President Harding.

Accordingly, among the younger members of the Dow chemical community the favorite anecdotes of the early days are those that now seem to be apocryphal: Dr. Dow riding bicycle races for a watermelon prize and playing Santa Claus, his hair and beard all white with corn-starch, handing out presents to the entire staff which gathered in the shack that served as laboratory, drafting room, and accounting office. Best of all they like the story of the tar roof. When the wooden buildings were roofed with tar paper, a regular maintenance job was to keep these coverings sound by frequent retarring. Watching a careless boy leisurely at this work, Dr. Dow instead of reprimanding him, invited him down from the roof for a practical demonstration of how such a job should be done. He himself climbed up and began laying the tar-brush about with such effect that within a few moments he had completely surrounded himself with a broad band of fresh tar. There he perched on the rooftree, marooned in a sticky sea from which he was only rescued with difficulty by three men, an additional ladder, and a long plank. Nobody laughed more heartily than did Dr. Dow himself, for he had a strong sense of humor and a thoroughly democratic attitude so that the closest personal intimacy grew up in the "Dow family".

Even during the first eight years when commercial progress lagged woefully behind technical improvements Dr. Dow had high courage and the assurance of success. His enthusiasm of faith he communicated to his associates by his trust in them, by encouraging rather than criticizing.

About 1900 the nucleus of the great Dow organization began to grow. From Case were recruited Thomas Griswold, Jr., and Edwin O. Barstow, respectively engineer-designer and operating chemist. These two and Dr. Dow used to argue hotly over ways and means, so that Earl Bennett, a Midland boy who was bookkeeper and shipping clerk, confesses that sometimes the rest of the little staff, huddled outside the laboratory door, feared that they would have to break up a pitched battle. But sooner or later, Dr. Dow's deep voice would break out in the dulcet strains of "Old Black Joe" and the battling trio would end their discussion in a roaring chorus. In 1906 Charles J. Strosacker came from Case to Midland to contribute mightily to the development of chloroform, carbon tetrachloride, carbon bisulfide, and during the War to the indigo process. From Case also came Mark Putnam to work on salicylates and later on aniline oil. From the University of Michigan was drawn another young organic chemist, William J. Hale, a dynamo of ideas, to whom belongs chief credit for the famous phenol process. In 1916, James T. Pardee, long a stockholder and director, came to help handle the fast-growing business end of the enterprise.

These are but a few of the group of good men Dr. Dow bound to himself, mentioned because they were longest and closest associated with him. His story, as he himself was forever insisting, is peculiarly the story of building up an organization. His was an original genius, stronger possibly in engineering than in chemistry, with a driving urge to find out if chemical operations might not be carried on in some new, shorter, cheaper way. Often he discarded little apparatus for large. Big units, as he was always pointing out, require no more control than small units, and far less control than a battery of small ones. When, in 1930, he was awarded the Perkin medal, his paper was a masterful

Dow Chemical plant, Aug. 31, 1900. Above shows rear of decomposer houses and in rear site of original bromine plant: below, to the left, is the lime house; and to the right, the power house and the warehouse.

The office force in the early days. In the first row, left to right, are: F. N. Lowry, E. W. Bennett, E. O. Cross, Fred Vance, G. Lee Camp; in the back row, Jim Smith, E. O. Barstow, J. E. LeFevre, and J. C. Graves.

exposition of the economics of the large unit and automatic control in chemical production.

Dr. Dow's courage and originality were tested in the commercial exploitation of his brine products. At first, as was then trade custom, Dow chemicals were sold through agents: Edward Hill in New York, Fred Clark in Cleveland, and Ellis Jackson in Philadelphia. In 1901 he was one of the first American chemical makers to send out his own salesmen; but this was not successful until 1904 when Rupert E. Paris became sales manager. He was succeeded in 1908 by W. H. Van Winkle, who left in 1917 to establish a sales agency of his own.

When Dow first began to market bromine the price was a dollar a pound, and although offered for half this price, leading American pharmaceutical houses refused to buy. They were importing refined salts from Germany and were loathe to undertake this additional manufacturing or to jeopardize their German connections. Accordingly, Dow went to work on the bromo-cyanide gold extraction process and finally contracted to supply large quantities of bromine to the Telluride Reduction Company of Colorado at 19.3¢ a pound. Well and plant capacity were greatly increased, but still Dow was unable to fill the rush of orders so down to Pomeroy, Ohio he went to purchase bromine at 25¢ to fulfill his obligations. Fearful over necessary supplies the Telluride people switched to another extraction reagent, cancelling their contract, and leaving Dow in a most uncomfortable position. It was then that Dr. Dow determined to manufacture refined bromine salts.

Despite a duty of 25 per cent., the German Bromine Cartel commanded the American market where they were selling potassium bromide at 40¢, the same price as in Germany. The first Dow shipment was to Japan and negotiations were proceeding towards contracts with two American pharmaceutical houses when there suddenly appeared in Midland one Herr Jacobsen, envoy of the Cartel. Dr. Dow was with his family en route to California; but he was recalled by telegram and the famous war declaration was delivered on neutral ground at a hotel in St. Louis.

Said the German envoy, "You are shipping bromides out of the United States, is it not so?"

"Yes," Dr. Dow admitted.

"That," exploded Jacobsen, "we cannot allow."

"And how do you propose to stop us?"

"For every pound that you export, we will import into the United States two pounds."

"Those arrangements, my dear sir, are entirely agreeable to the Dow Chemical Company, and I don't see that you and I have anything more to discuss."

Off stamped the German envoy and two days later his agents in New York flooded the trade with a circular letter pricing bromide at 20¢. Dr. Dow immediately decided to carry the war into the enemy's country. An agent in Germany was liberally stocked with American bromides and a price exactly 25 per cent. (the amount of the U. S. duty) below the Cartel's domestic quotation was announced.

For two years the battle raged, and then the Germans agreed to withdraw from price competition in the United States provided that Dr. Dow would stop shipments to Germany. Here again Dr. Dow put into practice his philosophy of chemical prices, for in 1909 he entered into ten year contracts with Mallinckrodt and Powers and Weightman based on the average price published in the trade papers. At that time this was 12½¢. When the World War started, domestic quotations ran up to $2.50 a pound, which at the agreed discount netted Dow $1.70 at a time exporting brokers in New York were bidding as high as $4. Throughout the War, not only did Dr. Dow live rigidly up to this contract; but he also materially increased the amounts which he shipped to his regular customers, enabling them to take advantage of the war export business.

Jokingly Dr. Dow attributed his scrupulous integrity to his New England ancestry. For although born in Bellevue, Ontario, February 26, 1866, both his parents were of old Connecticut families, and in that state, in the town of Birmingham, his younger boyhood was spent. His father, Joseph H. Dow had married Sarah Bunnell, whose father exerted a potent influence on the growing boy. Together, they roamed the woods on long tramps, collecting birds' eggs and arrow heads, minerals and butterflies. His love of flowers and trees was doubtless cultivated under his Grandfather Bunnell's guidance, for the old gentleman was a great gardener.

In later years horticulture became Dr. Dow's absorbing hobby. His estate of over a hundred acres with its Japanese and its modernistic gardens and its great apple orchards was not only famous for its beauty but also for the wonderful quality of its fruit. Such was the practical bent of his mind that even his relaxations must always be grounded upon some useful purpose. For exercise he delighted to plant trees and shrubbery, digging the holes himself, and transplanting to rearrange the plan of his gardens.

These interests he carried over into the plant in the manufacture of insecticides, and they reached out also to embrace the whole city. When he first moved there Midland was a muddy, ugly derelict of the lumber boom that had stripped the rolling landscape of its forests of white pine. By precept and example, by his liberal gifts and wise planning, he helped transform Midland into a notably beautiful community of homes with fine grounds and splendid parks. These horticultural triumphs he accomplished in a most unfriendly environment of sandy soil, hard winters, and short growing seasons.

To Midland he was always a practical benefactor. His assistance secured the Carnegie Library and his ideas made the Midland County Court House a building of rare artistic distinction. The Community Center was his direct gift and without his help the Country Club could never have been realized.

Keen as his public interests and absorbing as his industrial enterprises always were, Dr. Dow was a devoted husband and father. On November 16, 1892, he married a wife from Midland, Grace Ball, and they had seven children. Save one boy who died young, all these children grew up and married; Helen to Dr. William J. Hale; Ruth to Leland I. Doan; Willard, who has succeeded him as president of the Dow Chemical Company, to Martha Pratt; Alden, a successful architect, to Vada Bennett; Margaret to Harry Townsley, a physician of Ann Arbor, Michigan; and Dorothy to Anderson Arbury, a dentist of Ann Arbor.

While the Dow Chemical organization was never a family affair, and its founder was far too just and farsighted to countenance nepotism, nevertheless in the Midland community many close connections have been bound together by marriage. Thomas Griswold married Dr. Dow's own sister, Helen; and E. W. Bennett's daughter married Dr. Dow's second son. Two of his sons- in-law are with the company: the one contributed the important phenol process, the other is sales manager; both won their spurs without influence.

On October 15, 1930, in his sixty-fifth year, Herbert Henry Dow died at the Mayo Clinic, having failed to rally from a serious operation. Two days later the people of Midland paid their first fellow-citizen a very great, quite spontaneous tribute. It was a cold, drizzly afternoon. The special train bearing his body was over an hour late. Great throngs stood in the rain, neighbors and employees, friends all, waiting there patiently to welcome him home. Just as his body was being borne from the train a great gust of wind blew the gray, chilly mists away and through a rent in the lowering clouds the autumnal sunset suddenly flashed all scarlet and gold. From the sorrowful throng arose a great reverent sigh of thanksgiving. There, that, they felt, was the right welcome for this courageous fighter, this bold leader, this firm friend.

INDEX

INDEX

INDEX

INDEX